# TEXTBOOK OF
# PRACTICAL BOTANY

# TEXTBOOK
# OF
# PRACTICAL BOTANY

Volume - IV

(Anatomy, Embryology, Ecology, Physiology, Cytology)

*By*

Dr. S. Sundara Rajan

Principal

Reva Institute for Science and Technology Studies

R.T. Nagar

Bangalore – 560 024

ANMOL PUBLICATIONS PVT. LTD.

NEW DELHI - 110 002 (INDIA)

ANMOL PUBLICATIONS PVT. LTD.
4374/4B, Ansari Road, Daryaganj
New Delhi - 110 002
Ph.: 3261597, 3278000
Visit us: .www.anmolbooks.com

*Textbook of Practical Botany*


First Edition, 2003

ISBN 81-261-1359-6 (Set)



PRINTED IN INDIA

Published by J.L. Kumar for Anmol Publications Pvt. Ltd., New Delhi and Printed at Mehra Offset Press, Delhi.

# CONTENTS


CHAPTER 1 INTRODUCTION 1-19

PART ONE
PLANT ANATOMY

CHAPTER 2 STUDY OF PLANT CELL 20 - 27

CHAPTER 3 TISSUES 28 - 40

Meristematic tissues (28); Parenchyma (30); Collenchyma (32); Sclerenchyma (34); Xylem 357) Phloem (38)

CHAPTER 4 ANATOMY OF ROOTS 41 - 53

*Helianthus* (41); *Bean* (42); *Ranunculus* (43); *Tinospora* (44); *Prop root of Banyan* (45); *Smilax* (47); *Canna* (47); *Maize* (50); *Musa* (51); *Orchid root* (52)

CHAPTER 5 ANATOMY OF STEMS 54 - 68

*Helianthus* (54); *Tridax* (56); *Xanthium* (56); *Ricinus* (57); *Cucurbita* (59); *Tila* (61); *Maize* (62); *Vanda* (63); *Asparagus* (64); *Canna* (67)

CHAPTER 6 ANATOMY OF LEAVES 69 - 76

*Helianthus* (69); *Dalbergia* (70); *Nerium* (71); *Paddy* (72); *Maize* (73); *Stomata* (74)

CHAPTER 7 ANOMALOUS STRUCTURE OF STEMS 77 98

*Aristolochia* (77); *Tecoma* (78); *Amaranthus* (79); *Bougainvillea* (81); *Peperomia* (83); *Leptademia* (84); *Salvadora* (85); *Strychnos* (87); *Mirabilis* (88); *Argyreia* (89); *Chenopodium* (91); *Bignonia* (92); *Nyctanthes* (93); *Boerhavia* (95); *Dracena* (96); *Storage root of Beta* (97)

PART TWO
EMBRYOLOGY

CHAPTER 8 FLOWER AND ORGANOGENY 99 - 102

CHAPTER 9 ANTHER AND MALE GAMETOPHYTE 103- 107

CHAPTER 10 OVULE AND FEMALE GAMETOPHYTE 108 - 113

CHAPTER 11 ENDOSPERM 114-116

CHAPTER 12 EMBRYOGENY 117 - 120

## PART THREE

CHAPTER 13 ECOLOGY 121 - 163

**Study of Vegetation** - **1.** To determine the Minimum size of the quadrat by species Area-curve method (124); **2.** To determine the minimum number of quadrats to be laid down in the field (124); **3.** To determine the density, frequency and abundance of different species in an area (125); **4.** To Study the vegetation by line transect method and to determine the frequency of individual speicies (126); **5.** To study the vegetation by point frame method and to determine the frequency of individual speicies (126); **6.** To study the vegetation by Biological Spectrum Method (126); **7.** To determine the biomass of green plants in a given area (127);**7.** To determine the vegetational cover in an area (128); **8.** Study of Soil texture (129); **9.** Study of soil temperature (129); **10.** Measurement of soil pH (130); **11.**To determine the moisture content (%) in soil (130); **12.** Water holding capacity of soil (130); **13.** Calculation of carbonate content, nitrate content and base dificiency of soil samples (130); **14.** Test for the presence of Inorganic salts (131); **15.** Measurement of temperature, relative humidity and light intensity of air (atmoshphere) (131); **16.** Measurement of light intensity (131); **17.** Measurement of rainfall of a given area (132); **Ecological Groups - Hydrophytes** (133); *Hydrilla* (134); *Nymphaea* (135); *Trapa* (138); *Potomogeton* (140); *Eichhornia* (140); *Vallisnaria* (143); *Jussieua* (143); *Ceratoph*yllum(145); *Nelumbo* (145); *Pistia* (146); **Xerophytes** (146); *Causarina* (148); *Calotropis* (148); *Euphorbia* (150); *Asparagus* (153); *Cynodon* (155); *Muehlenbeckia* (157); *Ruscus* (157); **Halophytes** - *Rhizophora* (159); **Parasites** - *Viscun* (160); *Cassytha* (161); *Santalum* (161); *Custuca* (`61); *Orobanche* (161); **Epiphytes** - *Vanda* (162);

## PART FOUR

CHAPTER 14 PLANT PHYSIOLOGY 164 - 213

**Plant Cell** - **1.** Protoplasmic Streaming (168); **2.** Importance of Living Membrane (168); **Solutions, Suspensions and Colloids** - **3.** Preparation of 5% solution of sucrose (w/w) (169); **4.** Preparation of 5% solution of Sucrose (w1) (169); **5.** Preparation of 80% alcohol (v/v) (169); **6.** Preparation of 1 Molar solution of Glucose (169); **7.** Preparation of 1 molal solution of glucose (169); **8.** Preparation of 1 normal solution of NaoH (169); **9.** Preparation of 1 normal $H_2So_4$ (170); **10.** Preparation of a Suspension (170); **11.** Preparation of an Emulsion (170); **12.** Demonstration of Brownian Movement (170); **13.** Demonstration of Tyndall effect (171); **14.** Separation of Starch and Salt by dialysis (171); **15.** Demonstration of Diffusion (171); **16.** Demonstration of Imbibition (172); **16.** Demonstration of Imbition Pressure (172); **17.** Physical Demonstration of Osmosis (Thistle Funnel Experiment) (173); **18.** Physiological Demonstration of Osmosis (Potato Osmoscope) (173); **19.** Prepartions of a Semi permeable Membrane (Collodion bag) (174); **20.** Demonstration of plasmolysis (175); **21.**Determination of Osmotic Concentration of Cell sap by Plasmolytic Method (175); **22.** Measurement of Diffusion Pressure deficit (Water potential) (176); **23.** Comparison of concentrations of two solutions in relation to cell sap by turrgidity method (177); **24.** Development of turgor pressure due to osmosis (177); **25.** Demonstration of endosmosis and exosmosis Using boiled Eggs (178); **Absorption of Water** - **26.**

Absorption of water by roots (178); **Ascent of Sap - 27.** Root Pressure (179); **28.** Demonstration of Path of Ascent of Sap (180); **TranspIration - 29.** Demonstration of transpiration by bell Jar Method (180); **30.** Demonstration of unequal transpiration from the leaf surfaces by Cobalt chloride method (180); **31.** Comparison of rates of transpiration between upper and lower epidermis of leaf by bell jar Method (182); **32.** Unequal Transpiration in the lower and Upper epidermis - 4 leaves experiment (182); **33.** Comparison of rate of absorption and rate of Tranpiration (182); **34.** Development of suction force due to transpiration (183); **35.** Measurement of rate of transpiration by Ganong's potometer (184); **36.** Measurement of Transpiration by Farmers' Potometer (184); **37.** Demonstration of Continuity of air spaces in the leaf (185); **38.** Demonstration of Guttation (186); **39.** Measurement of Stomatal Opening by Darwin's Potometer (186); **Photosynthesis - 40.** Evolution of $O_2$ during Photosynthesis (187); **41.** Chlorophyll is necessary for photosynthresis (188); **42.** Light is necessary for Photosynthesis (Ganong's light screen experiment) (188); **43.** $CO_2$ is necessary for photosynthesis (Mohl's Half leaf experiment) (189); **44.** Effect of different wave lengths of light on photosynthesis (190); **45.** Measurement of Rate of Photosynthesis by Wilmott's Bubbler (190); **46.** Demonstration of presence of starch in chloroplasts (191); **47.** Study of photosynthesis by gass analysis (Ganong's Photosynthometer) (191); **48.** Extraction of Chlorophyll Pigments By Solvent wash method (193); **49.** Separation of leaf pigments by circular paper Chromotography (194); **50.** Separation of leaf pigments by Ascescending paper Chromatography (195); **51.** Demonstration of Hill's reaction in isolated Chloroplasts (Dye reduction method) (195); **Respiration - 52.** Demonstration of Anaerobic respiration (196); **53.** Demonstration of Fermentation (Kuhne's fermentation flask) (197); **54.** Release of $Co_2$ during aerobic respiration (conical flask method) (197); **55.** Evolution of heat during respiration (198); **56.** Demonstration of use of $O_2$ during Respiration (198); **57.** Demonstration of evolution of $CO_2$ during aerobic respiration by gas flow method (199); **58.** Evolution of $CO_2$ during an aerobic respiration (199); **59.** Measurement of Repiratory quotient (RQ) of different respiratory substances using Ganong's (200); **60.** Comparison of Respiration and Photosynthesis (201); **Enzymes - 61.** Test of Oxidase (202); **62.** Test for Peroxidase (202); **63.** Test for amylase (203); **Nitrogen Metabolism - 64.** Demonstration of nitrates in green plants (203); **65.** Experiment to demonstrate that atmospheric $N_2$ is not directly absorbed by green plants (203); **Growth and Hormones - 66.** Measurement of Growth by arc auxanometer (204); **67.** Measurement of growth by Pfeffer's Auxanometer (205); **68.** Measurement of growth by space markers (205); **69.** Effect of Auxins (Indole acetic acid - IAA) on plant growth (205); **70.** Effect of Gibberellins on plant Growth (206); **71.** Effect of Cytokinin on expansion of Cotyledons (206); **Plant Movements - 72.** Geotropism (Horizontal Klinostat) (206); **73.** Effect of Geotropism (Vertical Klinostat) (207); **74.** Phototropism (Heliotropism) (208); **Germination - 75.** Viability (Germinability) test for seeds (209); **Carbohydrates - 76.** Test for Cellulose (209); **77.** Test for Cutin (209); **78.** Test for Hemicellulose (209); **79.** Test for Suberin in the cell wall (210); **80.** Test for Lignin (210); **81.** Test for Mucilage (210); **82.** Test for Latex (210); **83.** Test for Glucose (Fehling's test) (210); **84.** Test for Glucose (Bendecit's test) (211); 85. Test for Sucrose (211); **86.** Test for Starch (211); **Proteins - 87.** Biuret Test (211); **88.** Xanthoproteic Test (211); **Fats - 89.** Test for Fat (212); **90.** Test for Fat (oil) (212); **91.** Test for presence of fatty acids by saponification (212); **92.** Test to Determine the saponification value of Fat (212); **Nucleic Acids** - (Microchemical test) **93.** Test for RNA and DNA (213);

## PART FIVE

**CHAPTER 15** **CYTOLOGY** **214 - 227**

Techniques of Cytological Study (214); Micrometry (219); Meiosis (221); Mitosis (225);

# Preface

Laboratory study is an integral part of the knowledge that a student acquires in the college. This is true not only for botany but for all science subjects. Theoretical study of a subject has to be supplemented with laboratory experiments to provide a comprehensive understanding.

One of the main advantages of laboratory or practical study is the first hand knowledge that a student acquires about the material of study.

A systematic, structured and disciplined method of work is necessary in the laboratory to obtain maximum benefit. In order to help the students realize their goal practical manuals indeed play an important role.

**Textbook of Practical Botany** aims to provide the students, a useful hand book to guide them in the proper study of the subject material. Theoretical details are kept at the minimum and more emphasis is laid on understanding the structural details to enable the student a proper identification. A large number of properly labelled original diagrams are incorporated in the book as they are integral to the understanding of the subject.

This work is organised in the following four volumes:

Volume One deals with Algae, Fungi, Bacteria, Lichens, Viruses and Microbiology.

Volume Two deals with Bryophytes, Pteridophytes and Gymnosperms.

Volume Three deals with Taxonomy and Economic Botany.

Volume Four deals with Anatomy and Embryology of Angiosperms, Ecology, Plant Physiology and Cytology.

I have taken utmost care in the presentation of matter, style of writing and choice of illustrations so that the volumes are user friendly. The examples chosen for study are such that the books should cater to the needs of all Indian Universities.

Constructive criticism and suggestions are most welcome.

I am thankful to Mr. J.L. Kumar, Managing Director of Anmol Publications Pvt. Ltd., New Delhi, for undertaking the publication of the book.

**Dr. S. Sundara Rajan**
Principal

Reva Institute for Science and Technology Studies
R.T. Nagar
Bangalore - 560 024

CHAPTER 1

# INTRODUCTION

The study of science subjects is carried out at two levels,viz., *Theoretical* and *Practical*. Theoretical study is carried out in lecture classes when the students are imparted all the details regarding a particular topic. This has to be supplemented by practical or Laboratory studies, when the students come into active contact with the objects about which they would have heard in lecture classes. Therefore it may be safely said that laboratory studies form an integral part of a comprehensive study of any topic in science subjects.

In the study of plants, laboratory assumes a special significance as the students acquire a meaningful and firsthand knowledge about the plants which they study in lecture classes. In fact as far as botany is considered any study of a plant without being supplemented by laboratory observation is futile. It is only in the laboratory study that the student can dissect, observe and analyze the various features of the plant and it is only a study of this type that makes the student acquire a meaningful knowledge about the pant. Besides observation and study in the laboratory plants are also to be studied in field to obtain an account of their habitat, habit, association and such other details.

In this chapter an attempt is made to explain some of the essentials that a student should understand before he begins a practical work.

***Cleanliness:*** This is the most important point that a student should keep in mind while in the laboratory. The working table and the surroundings should be kept dry and spotlessly clean.. All the requirements needed by the students such as petri dishes, slides, reagent bottles, stains etc. must be properly labeled and arranged neatly. The microscope and particularly the optical parts must be cleaned with a white lint free cloth. After the completion of the assigned work, the student should take care to clean the place, remove any debris of the plant material and put it into the dust bin.

The students should work in silence and independently unless the assigned experiment is to be carried out by a group. When a particular work is assigned to a student, he must carefully plan and divide the available time for experimentation, observation and recording of the results. Whenever a doubt arises, it is always better to consult the teacher in charge, than a classmate. It is only by a proper and systematic work, that a student can derive maximum benefit from the practical classes.

**Laboratory requirements**

The department provides certain amenities, instruments etc to each student for the purpose of practical work. But a student also should carry some of the requirements in order to be effective in the practical class.

One cardinal principle that every student should remember in a practical class is to have one's own requirements and not to depend on anyone. Never lend or borrow any requirements. Borrowing or lending a pencil, needle or blade will disturb the work of both.

In general every student should individually possess the following requirements

1. *Two sharp pointed needles* mounted on stainless steel or plastic handle. This is essential for the purpose of separating and teasing the microscopic material.
2. *A pair of forceps:* This is essential to remove the material either from the bottle or from petri dish.
3. A sharp properly honed razor to cut sections of the plant material.
4. A new safety razor blade. This will help either to cut section or to trim the material.
5. Two fine hair paint brushes. This will be of use in lifting the sections of plant (with out damage) from the petri dish on to the slide.
6. Two sharpened pencils (H, Hb or 2b).
7. A pencil eraser.
8. A pencil sharpener.
9. Glass droppers.
10. A piece of thin, white, lint free cloth to clean the optical parts of the microscope.
11. An observation book to note down immediately the points of interest.
12. A practical record.
13. Glass slides and cover slips(if they are not provided by the department).

**Scheme of Practical Work**

Practical work in the laboratory will be most useful and effective if the student keeps in mind the following:

1. Maintain silence during working. All the requirements for the practical work must be absolutely clean.
2. Every student should work independently, unless the experiment involves the participation of others.
3. Before beginning the practical work, listen to the instructions of the teacher attentively and work accordingly.
4. Follow a systematic procedure in a study.
5. If the work involved is the morphological study of an organism, follow the procedure given below:

   *(a) Study of external characters* - for example whether a plant is microscopic or macroscopic, unicellular or multicellular, filamentous or colonial etc.

   *(b) Study of internal structure* like details of cell, tissues etc. If the material requires sectioning, sections are to be cut and internal details should be observed.

   (c) Study the reproductive details like asexual and sexual reproductive organs.
6. Mcunt the material after properly teasing and separating the filaments, mycelia etc. Stain the material wherever necessary and cover the preparation with a cover glass.
7. Observe the slide carefully under the microscope, first under low power and then under high power.
8. Make sure that you have observed all the details as instructed by the teacher.
9. After all the observations are over, record your findings, structural details etc. in the practical records in the form of diagrams. Draw proportionate diagrams in the proper place on the drawing sheet. Diagrams should never be drawn to the very edge of the drawing sheet. Write suitable explanatory notes pertaining to the diagrams on the interleaf.

**Laboratory Equipments**

In addition to the individual requirements carried by the student, the department generally provides the following equipments/instruments etc.to the students. Sometimes these equipments may be given to each student .

These are :

(1) Students compound microscope .
(2) Dissection microscope.
(3) Binocular Research microscope.
(4) Watch glasses.
(5) Cavity blocks.
(6) Slides and cover glasses.
(7) Petri dishes.
(8) Enamel tray.
(9) Polythene/glass wash bottle.
(10) Spirit lamps.
(11) Glass marking pencil.
(12) Microtome.
(13) Staining /bottles.
(14) Stone/ for honing the razor.
(15) Chemicals and stains.
(16) Culture media(to be prepared by students).

In the list of requirements mentioned above, we will discuss some in detail.

**Microscope**

Of all the instruments used by students in the botany laboratory, the one that is most indispensable to the students of biology is the 'Microscope'. It is not an exaggeration to say that no biological study in the laboratory is possible without the microscope.

The function of the microscope is to enlarge the small objects, so that they become visible to the human eye. In reality the magnification power of the microscope lies in its resolving capacity between one particle and another. A microscope distinguishes one particle from another which are situated very closely. A human eye cannot resolve particles that are lying closer than 0.025 mm. In its function a microscope can be compared to a human eye, but much more effective than that.

There are different types of microscopes, but all of them are based on the same principle with variations in their construction. Some of the types of microscopes used in the laboratory are:

(1) Compound microscope.
(2) Binocular Research microscope.
(3) Dissection microscope.
(4) Stereoscopic dissection microscope.
(5) Phase contrast microscope.
(6) Electron microscope etc.

Of the six types of microscopes mentioned above the compound microscope and the dissection microscope are the ones that are most frequently used by students. The Electron microscope is very costly and a very highly sophisticated instrument.

In this book the compound microscope and the dissection microscope are explained in detail as the students should be familiar with them.

**Compound Microscope**

The microscope has a strong 'U' shaped or 'V' shaped base. It is also called the horse shoe shaped base. Attached to the base is a bent limb. This limb can be tilted in any direction as it is attached to the base with an inclination joint.

Attached to the limb is a stage at its lower end and a body tube at its upper end. The body tube is attached to the limb with the help of a rack and pinion so that it can move up and down. To help in the movement of the tube during focusing, there are 2 knobs viz.,*Coarse adjustment* knob and *Fine adjustment knob*.

The body tube at its top end consists of the ocular or eye piece lens. The eye pieces usually have three magnifications 5x,10x,and 15x, and can be used to get a low or high magnification. Attached to the lower end of the tube is a revolving nose piece with three objective lenses low (10x) medium (40x) and high (60x). Some times there will be a fourth objective of the magnification of 100x. This is also called the oil immersion lens (while viewing through an oil immersion lens a drop of oil-liquid paraffin has to be put on the viewing slide).

Below the objectives and attached to the limb is the *stage* of the microscope where the slides are kept for viewing. The stage may be circular, square or rectangular in shape. In order to keep the slides properly placed on the stage, there are two stage clips. In some of the costly microscopes, the stage has movable sliding system operated by knobs to move the slides. This is called the mechanical stage. The mechanical stage permits the movement of the slide backwards or forwards, left or right with the turning of the knobs provided for the purpose. A hole is provided in the stage to receive the light from below and illuminate the object. Attached to the limb, below the stage is a condenser (fixed or movable up and down) lens. Below the condenser lens is an Iris diaphragm to regulate the light passing through. Iris diaphragm is a metallic disc with an aperture. The aperture can be narrowed or widened by turning the lever. The condenser which is a system of lenses condenses the light rays passing through and converges them on the object.

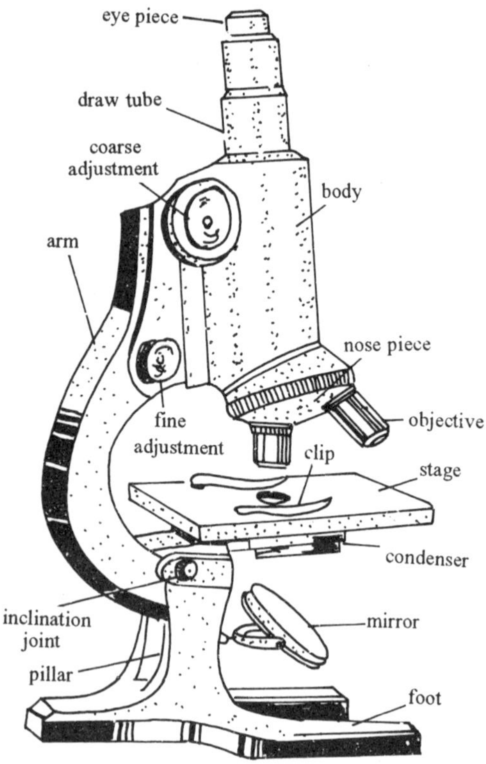


**Fig.1.1** A compound microscope

The source of illumination in a student's microscope is a mirror which reflects the natural light or electric light. The mirror has two faces - plain and concave and is attached to the limb below the condenser. The mirror is adjustable and can be turned to the direction of light. When day light is the source of illumination, the concave side of the mirror is usually used to converge the rays of light. In some of the costly microscopes, instead of the mirror there is a built in illumination ,This is a tube with a lamp (9v or 12v) attached to a transformer, which can be plugged to any AC source.

### Operation of the Microscope

Microscope should be placed in a well illuminated area but avoiding the direct light as it is harmful to the eyes. Generally northern light is preferable. When ' built in illumination' is employed, it is always better to use a blue or green filter to subdue the light.

The light has to be first focused on the condenser by adjusting the mirror towards the source of light. The amount of light entering the condenser may be regulated by closing or opening the iris diaphragm. The slide to be viewed is kept on the stage and fixed securely with the help of the stage clips. The object is first observed with a low power objective and then turned on to medium or high power, the specimen or section has to be covered with a cover slip. The slide should be observed with both the eyes open even while observing with monocular microscope.

**Care of the Microscope**

1. Clean all the parts of the microscope before and after use. The lenses and the mirror are to be cleaned with a lint free cloth or tissue paper.

2. Do not pull the microscope to the very edge of the table.

3. While lifting the microscope, use both hands. With one hand hold the base and with the other hold the arm.

4. The slide to be observed must be clean and dry ( specially the under surface of the slide).

5. Do not remove the eyes piece or objectives from the microscope.

6. The microscope should be covered when not in use.

**Binocular Research Microscope**

This is like the ordinary compound microscope except that, it has binocular head for providing a natural stereoscopic view.

**Dissection Microscope**

This is a very simple microscope with only one set of lenses. Microscopes of this type will not give a very high magnification, but are very useful in dissecting flowers, plant parts etc.

The microscope has a 'V' shaped base and a limb. The stage is made up of a transparent glass plate. Below the stage is a mirror to focus the light. Attached to the limb is a mechanical arm which can be moved up and down with the help of a rack and pinion movement. The arm can also be moved horizontally. Attached to the free end of the arm is a holder to accommodate the lens (10x,15x etc.).

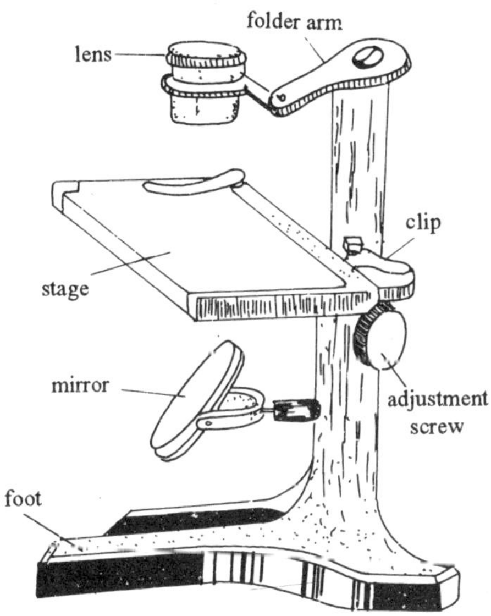


**Fig.1.2** A dissecting microscope

**Operation:** The specimen to be observed / dissected is placed on the stage and located with the help of lens. Floral parts or others may be then dissected with the help of two needles, while still observing the specimen.

**Stereoscopic Binocular Dissection Microscope**

This is a combination of compound microscope and dissection microscope. While it has the higher magnification power of a compound microscope with two sets of lenses, it has retained the versatility and the manoeuvourability of a dissection microscope.

Phase contrast and Electron microscopes are not normally used by students.

**Other Laboratory Equipments**

**Stains:** These are colouring fluids. The specimens to be observed (algal or fungal material or sections of plant parts, would be generally decolourised due to preservation. These are to be coloured in order to observe them clearly. The following are the stains that are generally used in the laboratory.

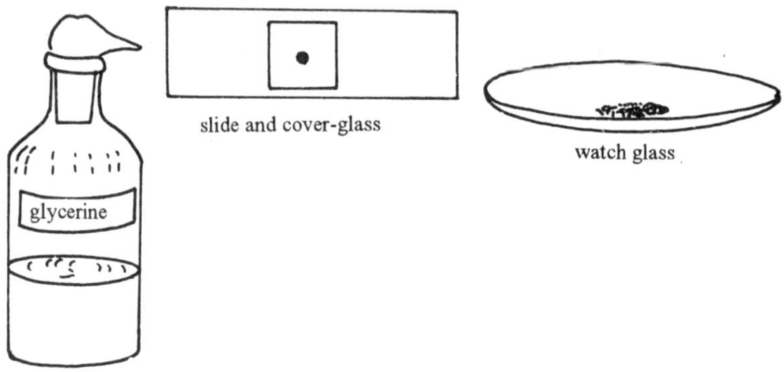


**Fig.1.3** Dropping bottle

**(a) Saffranin:** This is the most commonly used stain. (1) It stains nucleus and lignified cell walls.
Preparation —

Saffranin —1gm.
95% (1) Ethyl alcohol –50ml.
Distilled water – 50ml.
An aqueous stain may also be prepared.
Saffranin – 18ml
Distilled water – 100ml.

**(b) Fast Green**

Fast green – 1gm.
90% Ethyl Alcohol – 100ml.

This stain is used to colour parenchyma cells and cytoplasm. Fast green may also be prepared in clove oil ( 75ml clove oil+25ml 100% alcohol).

(c) **Cotton blue.**

Cotton blue –1gm.
Lactophenol – 100ml.

**(d) Haematoxylin**

The dye is obtained from the heartwood of the tree *Hametoxylon campechianum,* There are two methods of preparation of the stain. These are:

1. *Heidenhain's Hametoxylin*

Dissolve 0.5 or 1.0 gm of the powder in 100 ml of distilled water. Filter and store in a dark bottle away from light. The stain is ready for use after 4-5 days of ripening. Ripening is not necessary if the stain is prepared in 70% ethyl alcohol.

2. *Delafelds Haematoxylin*

Sol.1 Aqueous solution of ferric ammonium sulphate 100ml).

Sol.2 One gm dye dissolved in 6ml of absolute alcohol.

mix solutions 1 and 2. To this mixture add 25 ml of glycerine and 25 ml absolute alcohol. The solution should be allowed to stand until it becomes dark red.

**(e) Anilin blue**

Also called cotton blue or China blue. It is prepared as follows:

Aniline blue –1gm.

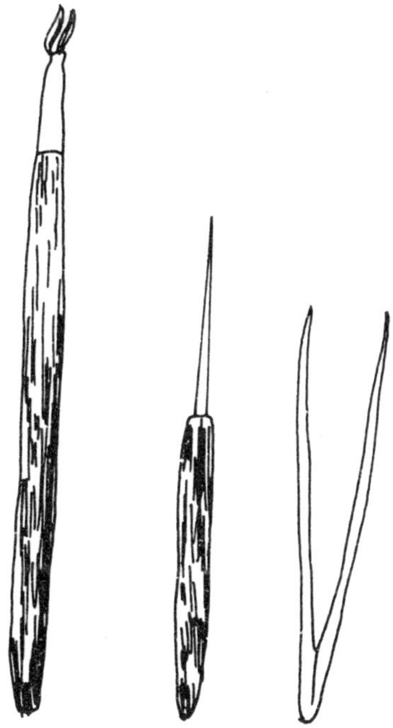

Fig.1.4 Brush Needle Forceps

Ethyl alcohol (90%)
or — 100ml.
Distilled water
Hydrochloric acid – 2drops.

**(f) Gentian Violet** (Crystal Violet)

Gentian Violet – 1gm.
90% Ethyl Alcohol 100ml.

**(g) Erythrosine**

Erythrosine – 1gm.
90% Ethyl alcohol – 100ml.

Erythrosine may also be prepared in clove oil (Absolute alcohol 0.5 ml +95ml Clove oil).

**(h) Light green**

Light green – 1gm.
Clove oil – 95ml.
Absolute alcohol – 5ml.

Light green may also be prepared without clove oil (in 100ml Absolute alcohol).

**( i ) Acetocarmine:** The stain is prepared as follows:

Add 1 gm of stain to 100 ml of boiling acetic acid (45%)or propionic acid. Cool and decant, A few drops of saturated ferric acetate solution is added and the mixture is cooled over ice. The solution is allowed to stand for about 12 hours. Filter and store in a dark coloured bottle under refrigeration.

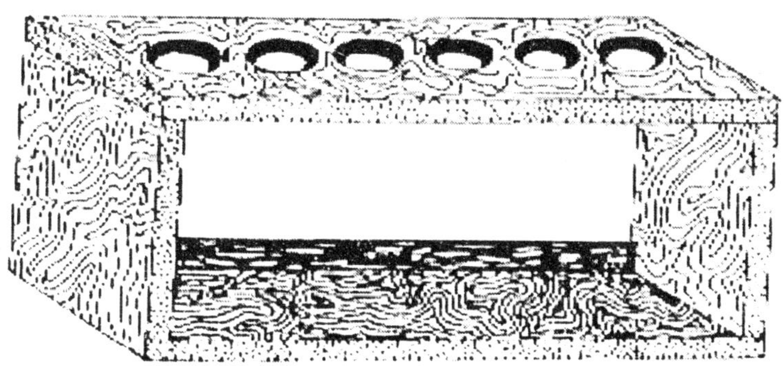

**Fig.1.5** Staining rack

**Specificity of Stains**

Usually specific stains are employed in colouring different plant parts. The stains are used singly or in combination. The most popular stain combination are:

Saffranin - fast green.
Haematoxylin - Erythrosin.
Haematoxylin - light green.

While stain combinations are employed to study pteridophytes, gymnosperms and Angiosperms, single stains are used for Avascular cryptogams.

Green or Saffranin for Algae and Bryophytes.

Cotton blue for fungi.

**The following table gives the specificity of stains to different plant parts or cell organelles.**

| Structure | Stain |
|---|---|
| Cellulose | Anilin blue<br>Haematoxylin<br>Fast green<br>Light green |
| Cutinized cell wall | Gentian violet<br>Erythrosine<br>Saffranin |
| Lignified cell wall | Gentian violet<br>Saffranin |
| Suberised cell wall | Saffranin |
| Cytopalsm | Fast green<br>Light green<br>Aniline blue |
| Chitin | Saffranin |
| Callose | Aniline blue |
| Achromatic figure | Aniline blue<br>Erythrosine |
| Proteins | Saffranin |
| Nucleus | Heamatoxylin<br>Saffranin<br>Gentian violet |
| Chromonemata | Heamatoxylin |

| | |
|---|---|
| Chromosomes | Saffranin |
| Plastids | Gentian violet<br>Haematoxylin |
| Mitochondria | Gentian violet |
| Flagella | Cotton blue |

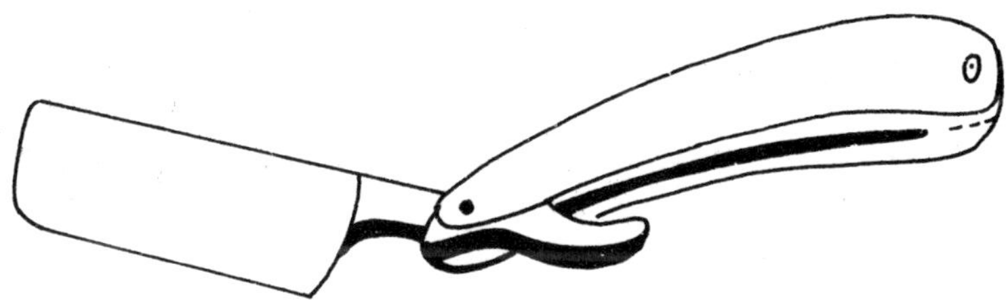

**Fig.1.6** Razor

**Fixing / Preserving Fluids**

The plants or plant parts collected in nature are to be killed and preserved in the laboratory for the purpose of storage and study.

In killing the specimen, the various life processes are stopped at that instant, so that after preparation for microscopic study various stages may be observed.

When ever a material is collected, it is better to kill it in the field, so that there will be minimum or no distortion. Alternatively the material collected in the plastic bags or bottles should be brought to the laboratory and killed / preserved. In preservation, the material must be completely immersed in the fluid. The following are some of the killing/preservative fluids generally used:

1. *Formaline acetic alcohol (F.A.A)*

This is the most commonly used killing /preserving agent. It is popularly called standard preservative.

| | |
|---|---|
| Ethyl alcohol (70%) | - 90ml. |
| Glacial acetic acid | - 5ml. |
| Formalin | - 5ml |

2. *Formalin - Propiono - Alcohol*

| | |
|---|---|
| Formalin | - 5ml |
| Propionic acid | - 5ml. |
| Ethyl alcohol 70% | - 90ml. |

3. *Carnoy's Fluid - I*

| | |
|---|---|
| Ethyl alcohol | - 30ml |
| Glacial acetic acid | - 10ml |

4. *Carony's Fluid - II*

| | |
|---|---|
| Ethyl alcohol | - 30ml. |
| Glacial acetic acid | - 30ml. |
| Chloroform | - 15ml. |

Carony's fluid is widely used in killing root tips, flower buds etc. for the purpose of cytological study.

5. *Randolph's Modified Navashin Fluid*

This consists of two solutions A and B.

*Solution A*

| | |
|---|---|
| Chromic acid | - 5.0 gms. |
| Glacial acetic acid | - 50ml |
| Distilled water | - 320ml. |

*Solution B*

| | |
|---|---|
| Commercial formalin | - 200ml. |
| Distilled water | - 175 ml. |
| Saponin | - 3gms. |

Mix A and B in equal volumes before use.

***Mounting media***

These are chemicals used for the purpose of mounting (Temporary/Permanent) the specimens on the slide for microscopic observation. The following are some of the mounting media.

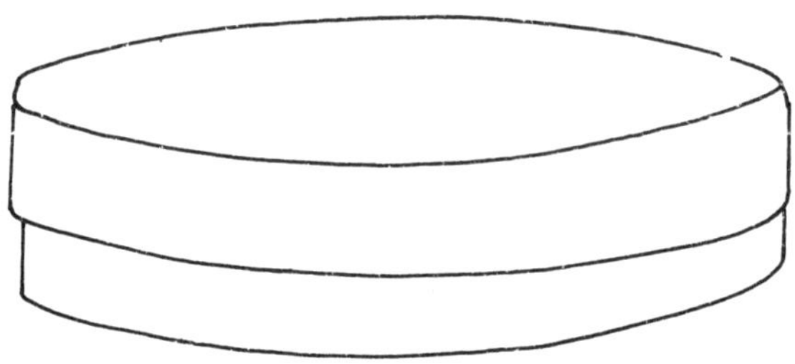

**Fig.1.7** Petri dish

1. ***Acid water***

HCl – 10-15 drops

water – 100 ml.

2. ***Lactophenol***

| | |
|---|---|
| Phenol | -100gms |
| Lactic acid | -100 ml. |
| Glycerine | -100 ml. |
| Distilled water | -100ml. |

3. ***Glycerine***

| | |
|---|---|
| Glycerine | - 15-20 drops |
| Formaline | - 1-2 drops |
| Distilled water | - 100ml. |

4. ***Canada Balsam:*** This is a gum obtained from Gymnosperms. This fluid is readily available in the market.

5. ***Euparol:*** This is available in the ready to use form in the market.

Canada balsam and Euparol are used for permanent mounting, while the rest are used for temporary preparations.

***Staining rack:*** This is a wooden rack used for holding the staining bottles.

***Dropping bottles:*** This is a glass bottle having a narrow mouth fitted with a stopper. The stopper is grooved and has a beak. When the bottle is upturned the stain/fluid comes out in drops.

***Slides:*** These are readily available in the market. The usual size 25mm x 75mm. The thickness is about 1mm. Special types of slides called cavity slides are available for specific study.

**Cover slips** (Cover glasses)

These are meant to cover the materials when the preparation is ready for observation. The coverslips may be round, square or rectangular.

**Methods of Study of Plant Material**

The following is the procedure for a scientific study in the laboratory.

***Morphological study***

The specimen given to study external detail should be observed either with or without a magnifying glass. Whatever details can be seen are to be recorded. If the specimen requires microscopic study, it should be suitably prepared for the purpose of microscopic observation. There are two types of microscopic preparation.

***1. Whole Mount Technique***

This is suitable for the purpose of external study of algal and fungal material. Take a little of the material, stain it with a suitable stain (staining procedure is explained later.) Wash the excess stain in water / alcohol. Mount in a suitable medium (temporary or permanent) and observe under the microscope.

***2.Section cutting***

For the purpose of anatomical study, sections are to be cut of plant parts. In algae, globule of *Chara*, stolons of *Caulerpa*., receptacles of *Sargassum* etc., can be cut. Similarly fruiting bodies of fungi, thalli of bryophytes also many be cut in a suitable plane. In higher plants cylindrical plant parts like root, stem, petiole etc. may be cut for the study of internal structure.

***Planes of Cutting***: In order to study the internal structure from various angles, sections are cut in various planes.

***Transverse :***The section is cut at right angles to the longitudinal axis.

***Longitudinal :*** The section is cut parallel to the long axis (right angles to the transverse axis. The longitudinal section may be cut along a radius (Radial longitudinal section-RLS) or along a tangent ( Tangential longitudinal section-TLS).

Cylindrical organs like roots and stems are cut transversely or longitudinally, while dorsiventral organs like thalli of bryophytes, leaves etc. are usually cut in vertical transverse plane.

**Types of Section cutting**

There are usually 2 types of section cutting methods employed for plant parts. These are (a)Free hand sectioning and (b) Microtome sectioning

**(a) Free hand Sectioning**

For the purpose of laboratory study where slides are not made permanent, usually free hand sectioning method is employed. Free hand sections are cut mostly with the help of a good sharp razor (sometimes safety razor blades also may be used).

**Honing the razor** : The razor used for sectioning must have a sharp, even edge. There should be no nicks. Before cutting sections, the razor edge has to be honed to maximum sharpness either on a stone or on a leather strap. Oblique strokes, with the edge forward are to be given on the sone or strap. Usually oil drops are used on the honing medium to give maximum sharpness.

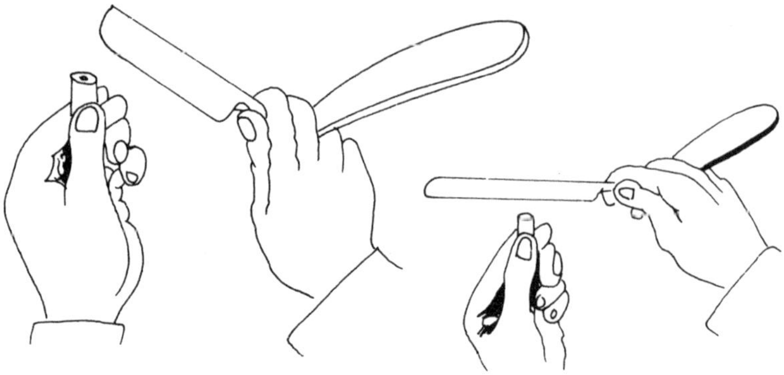

**Fig.1.8** Method of holding pith — Methoding of holding razor — Method of section cutting

**Method of cutting :** Open out the razor and bend the handle backwards. Hold the razor between the index finger and thumb. The handle should be free. The remaining fingers of the hand will be touching the back edge of the razor blade.

In order to cut a transverse section of a cylindrical part like stem or root, follow the procedure given below.

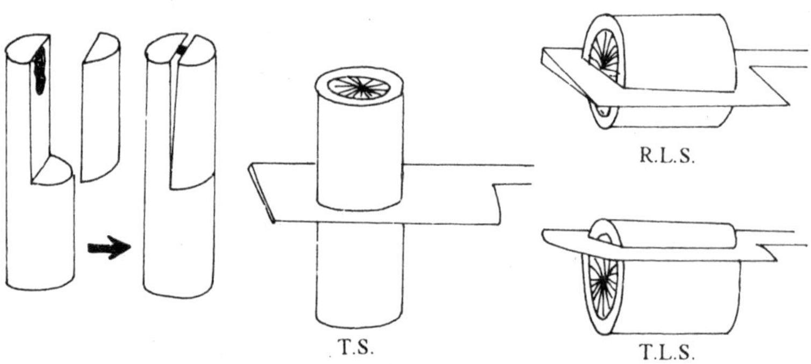


**Fig.1.9** Method of using pith for insertion of material

The stem or root should be cut into a 2-3" long piece. It should be held in the hand in an erect position between the thumb and index finger. The material held in the hand should be exactly at right angles to the razor's edge. The thumb holding the specimen should be slightly at a lower level. Flood the edge of the razor with water. Now using the index finger as the platform move the razor quickly several strokes across the material. During this process sections are cut and they will be collected on the blade of the razor. The sections now may be carefully transferred with the help of a fine brush in to a petridish containing water.

For cutting sections of delicate and fragile materials like thalli or leaves, a suitable supporting medium has to be used. For this purpose a pith, carrot or potato may be used. These are to be cut into rectangular cubes and then split at one end. The material for sectioning should be inserted in the slit and sections are cut

After section cutting the razor should be wiped dry. It should be greased and encased.

**Microtome Sectioning:** This is generally employed very frequently in the study of bryophytes, pteridophytes and other higher plants. The microtome sectioning is a long and complicated process. It takes anywhere between 15-20 days to process a material and to obtain the sections.

Microtome sectioning involves the following steps:

1. Killing and preserving the material.
2. Dehydration and infilteration.
3. Paraffin embedding .
4. Section Cutting.

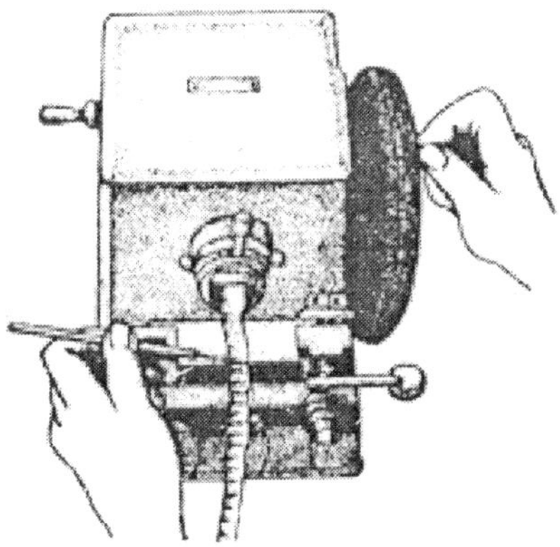

**Fig.1.10** Rotary microtome

The material like stem, leaf, flower etc. is to be killed and preserved in a suitable fluid.

(a) Cut the material into bits of suitable size.

(b) Place the material in small vials containing a fixative or preservative.

(c) Label the vial giving the details of place of collection , name of the material, preservative used and date of collection.

The material should be in the killing fluid at least for six hours (preferably a day).

**Dehydration:** In order to dehydrate the material, remove it from the preservative and wash thoroughly in water and transfer it to a suitable tube or vial.

Dehydrate the material in alcohol series (Ethyl alcohol ) as follows

| | | |
|---|---|---|
| Stage 1 | 50% Alcohol | First day |
| Stage 2 | 70% Alcohol | Second day |
| Stage 3 | 90% Alcohol | Third day |
| Stage 4 | 100% Alcohol | Fourth day |
| Stage 5 | Absolute alcohol and xylol (50/50) | Fifth day |
| Stage 6 | Absolute alcohol and xylol (25/75) | Sixth day |
| Stage 7 | Xylol | Seventh day |

When the material is transferred to pure xylol, there should be no fogging (i.e the solution should not turn milky). If there is fogging repeat the procedure starting from 4th stage.

The material in xylol will be completely dehydrated and looks transparent. It is now ready for infilteration by paraffin wax. The purpose of infilteration is to see that the wax enter into the material completely so that it can easily be cut.

**Infilteration:** Add a few pieces of paraffin wax everyday for about 4-5 days and then transfer the vial into a hot air oven, while temperature is adjusted to the melting point of wax. Everyday add a few pieces of wax. Adding of wax should be continued until all the xylol in the vial is replaced by paraffin wax. When the tube is smelt, there should be no smell of xylol. The material is now ready for embedding.

**Paraffin embedding:** In order to cut the material, it has to be embedded into a paraffin block. Prepare a paper boat or take a porcelain rectangular dish and smear glycerine to the inner surfaces. Pour molten paraffin into the porcelain dish and empty the contents of the vial into the dish. Before the paraffin solidifies, arrange the material in proper rows and in such a manner as desired. Leave it for cooling for 6-8 hours. For quick cooling the dish may be transferred to a bowl containing ice water. Carefully remove the solidified paraffin block from the dish and cut it into suitable pieces. The material is now ready for sectioning.

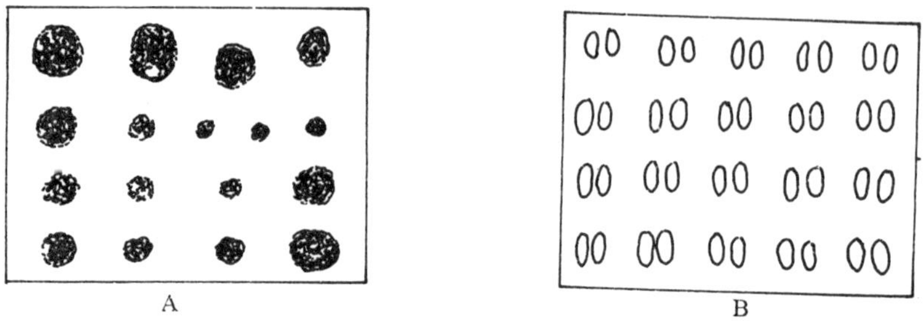


**Fig.1.12** Arrangement of materials in paraffin blocks - A) Flower buds in groups B) flower buds in pairs

**Sectioning:** Sectioning is carried out with the help of an instrument called microtome. There are various types of microtomes like Rotary, sliding, Freezing etc. The one that is normally used is a rotary microtome. The following procedure can be adopted in the case of the rotary microtome.

1. The paraffin block is to be cut into small pieces each having the embedded material.

2. The pieces of paraffin containing the material is to be fixed to wooden 'riders'. For this purpose, heat the top of the rider and the base of the paraffin piece with a hot knife and fix. Allow it to cool. The paraffin piece is now firmly fixed to the rider. In order to make the edges of the paraffin even, cut the edges and make them parallel.

3. Fix the wooden rider into the microtome so that the paraffin piece with the material is exactly parallel to the edge of the microtome knife.

4. Adjust the knob of the microtome to give the required thickness to the sections.

5. Rotate the wheel of the microtome in a clockwise direction. The cylinder containing the rider moves up and down on the knife edge. Sections are cut in a sequence and get attached to each other forming a paraffin ribbon.

6. Keep a needle in your left hand to hold up the ribbon as it lengthens. When the ribbon is sufficiently long, cut it and transfer it on to a black cardboard. Cut the ribbons into suitable length and transfer to the slide.

7. Smear the slides with an adhesive like egg albumen or gelatin and rub it well.

8. Pour some water onto the slide with the help of dropping bottle.

9. Transfer the paraffin ribbons on to the slide and drain excess water.Retain a little water on the slide.

10. Place the slide on a slide warmer. The ribbon expands, and so also the sections of the material. Now draw out the water completely and allow the slides to cool.

11. The slides should be allowed to stand for atleast 24 hours whereupon the sections will stick to the slide permanently.

The slides are now ready for staining.

**Staining procedure**

The plant material (whole mount or sections) are to be stained and mounted properly before they are ready for microscopic observations.

Algal filaments, fungal mycelia, pollen grains and spores, prothalli of pteridophytes are usually stained with a single stain - saffranin or green. In some case Aniline blue is used to stain the filaments of Algae and mycelia of Fungi. The staining procedure is as follows.

Place the material either in a watch glass or cavity slide and add a few drops of stain. The period of staining varies with different materials. But usually a 5-10 minute period is sufficient. Drain out the stain and observe the material under the microscope. If the material has taken less stain, it may be kept in stain for some more period, or if it has taken excess stain it can be removed by washing either with water or 50% ethyl alcohol.

After the material is suitably stained pass it through a series of alcohol grades to dehydrate it. Usually different grades of alcohol (70%,90%,100%) taken in a filler are put on the material. After passing through absolute alcohol the material is kept in xylol and it is now ready for mounting.

Fungal mycelia can be stained with aniline blue. Usually the material taken on a clean slide is stained with a drop or two of the stain prepared in lactophenol. The material can bemounted straightway as lactophenol is also a mounting medium. If the stain is not prepared in lactophenol the material may be transferred to fresh lactophenol and mounted.

**Double Staining Procedure**

In the case of most of the algae and fungi there is little or no tissue differentiation and hence single staining is sufficient. In some cases of plants with relatively more tissue differentiation, a single stain is not sufficient to mark out the various tissues clearly. In such cases 2-3 stain combinations are used to stain the tissues. For example lignified and non-lignified tissues are stained with a combination of acidic and basic dyes.This type of staining is also called differential staining where different tissues take up different stains.

The following are some of the popular stain combinations.

| *Main stain* | *Counter stain* |
|---|---|
| Haematoxylin | Saffranin |
| Haematoxylin | Erythrosine |
| Haematoxylin | Eosin |
| Haematoxylin | Orange G. |
| Saffranin | Fast green |
| Saffranin | Light green |
| Saffranin | Aniline bluc |
| Saffranin | Crystal violet |
| Crystal violet | Erythrosine |

***Staining Procedure for free hand sections***

**(a) Procedure for temporary preparations**

Free hand sections of root, stem etc., should be washed thoroughly in distilled water in a watch glass. After repeated washing the sections should be transferred to a watch glass containg the main stain. The sections should be kept in the stain for a period ranging from 30-40 minutes. After this period, the sections should be transfereed to a watch glass containing distilled water. The sections should be washed for 2-3 minutes whereupon excess of stain is removed. They should be observed under the microscope. If the main stain is saffranin only lignifed tissues should retain the stain, others should be destained. If the main stain is specific to non lignified tissues, only they should retain the stain. Then the sections should be washed thoroughly in water and repeatedly observed under the microscope. The sections can be mildly stroked with a fine hair brush, so that excess of stain comes out with distilled water. The sections are now ready for counter-staining. They should be transferred to a watch glass containing the counter stain (fast green etc). Counterstain acts quickly on the tissues. Hence one to two minutes of staining should be sufficient. The sections are now observed under the microscope. If there is excess of counter stain, it can be removed by washing sections with 15-20%glycerine. After the final stianing the tissues whould be demarcated clearly with no mixing of the stains. The sections are now ready for mounting.

**(b) Procedure for permanent preparaions**

If the sections of plant materials are to be mounted and made into permanent slides, the procedure should be slightly altered so as to include dehydration and clearing with suitable agents. The procedure is as fallows:

1.Stain with the principle stain (Saffranin, Haemotoxylin etc).

2.Allow the sections to remain in stain for about 15-20 minutes.

3.Transfer the sections to water and wash thoroughly. The sections should be stroked mildly until no more stain comes out, and the water should remain clear. If the main stain like saffranin is produced in 50% alcohol then destaining (washing) should be done in 50% alcohol.

4.The sections should be now dehydrated in alcohol grades of 50%, 70% and 100%.

5.In each grade of alcohol keep the section for about 5 minutes.

6.Transfer the sections from one grade of alcohol to the other with the help of a fine brush. Never use needles as they injure the sections.

7.Transfer the sections to the counter stain (fast green etc). The counter stains are usually prepared in 90% alcohol.

8.Keep the section in counter stain for about a minute. Wash the sections in 90% alcohol and transfer them to 100% alcohol.

8a. In some case counter stains are prepared in clove oil. In such cases counter staining should be done after the sections are completely dehydrated with absolute alcohol.

1. Sections are transferred completely dehydrated to a solution of 50:50 Absolute alcohol and xylol followed by 25:75 Absolute alcohol and xylol and ultimately the sections are transfered to pure xylol. Now xylol should not turn milky or turbid. If it turns milky, dehydrate the sections again through combination of absolute alcohol : xylol series.

The sections are now ready for mounting.

**Staining procedure for microtome sections**

The slides with the paraffin ribbons adhering to them have sections of plant material.The processing and staining is as follows:

1. Immerse the slides in a coupling Jar containing xylol. Leave them for 10 minutes . The paraffin dissolves in xylol and only the sections remain on the slides.
2. The slides are now kept in 50:50 absolute alcohol and xylol.
3. Pass the slides through a down grade series of alcohol. 100%, 90% and 70%
4. Transfer the slides to pure water and wash them thoroughly to remove traces of alcohol.
5. Keep the slides in a mordant like 4% iron alum. This is necessary only when Haematoxylin is used. After about 10 minutes wash the slides in water.
6. Transfer the slides to the main stain Haematoxylin. Saffranin etc. and keep them for 15-30 minutes.
7. Remove the slides from stain and wash in water.
8. Observe the slides under microscope. Usually the sections are overstained.
9. If the main stain used in Hametoxylin, destaining can be done with 2% Iron alum or with water containing few drops of concentrated HCI. Repetaedly observe the slides under the microscope. Stop destaining when only the nuclei have retained the stain.

**Caution:** When destaining is done with HCI, not more than 5-6 drops should be used for 75ml of water.

Excess acid will eat away the adhesive and the sections come out of the slide.

10.Wash the slides in water overnight if the destaining agent is Iron alum, otherwise washing for 30 minutes is enough.

11.Now pass the slides through an upgrade series of alcohol (5 minutes in each grade) 70% 90% 100%

12. When the sections are completely dehydrated stain them with the counter stain (Fast green etc) for a minute or two. Wash the excess stain in clove oil and transfer the slides to pure xylol. The slides are now ready for mounting.

**Mounting**

(a) *Mounting Media*

Mounting media are of two categories: Temporary mounting media and permanent mounting media.

A temporary mounting medium may be just water or water mixed with glycerine. If the temporary sections are to be retained for 2-3 hours, dilute glycerine (15-20%) is recommended.

Permanent mounting media are used when the slides are to be retained indefinitely. The following are some of the permanent media.

(1) **Canada Balsam:** This is the most widely used mounting medium in botanical preparations. It is available in ready to use bottles. It is a gummy substance. If the solution is very thick add little xylol.

Canada balsam is a resinous product of a Gymnosperm, Abies balsamea. The gum has very good refractive index and as such it is highly reccommended.

(2) **Lactophenol:** Lactophenol can be prepared in the laboratorty. Mix equal quantites of phenol cystals, lactic acid glycerine and distilled water.

(3) **Glycerine gelly:B** This is prepared as follows:

Gelatine 1 part.
Water 6 parts.
Glycerine 7 parts.

Glycerine jelly is suitable even for mounting fresh specimens (even without dehyderation).

Add gelatine to water, stir thoroughly and warm. Add glycerine and cool the solution. After sometime the solution settles into a jelly.

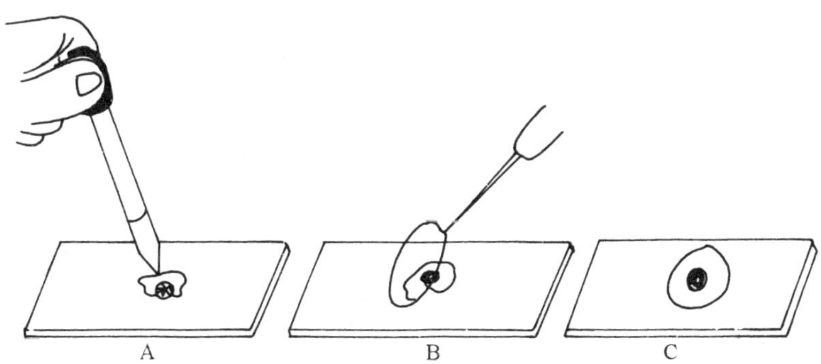


**Fig.1.13** Method of mounting the material

**(b) Mounting Method**

The sections (free hand) are to be mounted on to clean slide in a specific mounting medium. If the sections are microtome preparations they would be already adhering to the slide. In any case the sections are to be covered with a thin coverslip or cover glass which are readily available in the market. They may be round, square or rectangular. Select the one that is suitable for your preparation.

The sections or objects should be mounted on the centre of the slide. In order to ensure placing of the object on the centre of the slide, the following procedure may be adopted. On a white paper draw a rectangle of the same size as the slide. Join the opposite corners by drawing diagonal lines. The centre of the slide lies where the two lines cross. Place the slide on the rectangle and material or objects may be kept on the central point.

Place a drop or two of the mounting medium. Hold the coverslip at the edges (see the figure) and allow it to touch the edge of the mounting medium at an angle of approximately 45°. Slowly lower the cover glass on the mounting medium with the help of a needle. Take care to see that no air bubbles enter the mounting medium. If there are air bubbles, lift the cover glass and mount again.

Excess mounting medium may be carefully wiped with the help of a clean tissue. Allow the slides to dry. If necessary the slides may be transferred to a hot air oven for drying. The slides should be labelled at one end as shown in the figure. They are now ready for microscopic observation.

**Ringing of Slides**

Temporary preparations may be kept for few days if the borders are sealed. This procedure is known as ringing . Ringing is done by applying wax, canada balsam or even nail polish to the edges. Ringing may be done manually or by using a ringing table having a rotary metallic disc.

**Record of Practical Work**

Practical work carried out in the laboratory should be recorded. The botany practicval record book has two types of sheets viz., Interleaves and drawing sheets. Interleaves are meant to write the notes pertaining to the diagrams. The drawing sheets which are thicker than the interleaves are used for diagrams. The following points may be useful in maintaining a good record.

1. Use good pencils (HHB) with sharp points and a good quality eraser.
2. Leave a minimum of one inch margin on all the sides.
3. Always write diagrams of only one plant in a page. Do not mix up diagrams of other plants.
4. On the left hand corner of the drawing sheet write the date of the experiment.
5. Give the complete systematic position of the member on the top of the drawing sheet.
6. For microscopic organisms draw first the habit sketch followed by low power and high power view/s. High power view should show the cellular details.
7. In the case of anatomical drawings, draw the ground plant first, followed by a sector enlargement to show the cellular details.
8. All parts of the diagrams must be properly labelled.
9. Stages of reproduction, if any, must be drawn after the details of vegetative structure.
10. The description corresponding to the diagrams must be written on the opposite interleaf.

**Field Collection of plants**

Specimens of algae, fungi and bryophytes can be collected in the field for the purpose of laboratory study. Field trip is an integral part of botanical study and all students must participate in it. It is in the field study that the students will come to understand the natural growth, habitat and environmental association of plants. During a field trip to collect specimens, the following items are to be carried:

1. Pocket lens.
2. Fixative F.A.A or Carnoy's fluid.
3. Collection bottles.
4. Blotting paper.
5. Polythene bags
6. Field note book
7. Labels.

When the specimens are collected they can be fixed and preserved for the purpose of laboratory study. Use the required preservative and immerse the specimens in it. Label the bottle showing details of

Date of collection

Location

Preservative fluid (name) used.

Name of the plant

As it is necessary for the purpose of identification, details of habitat, associations with other plants if any, colour of the fresh specimen etc. should be recorded in the field note book.

# PART ONE
# ANATOMY

CHAPTER 2

# STUDY OF PLANT CELL

**Purpose :** To study the structure of a typical plant cell under microscope by using thin layer of onion peel.

**Materials :** Onion bulbs, slides, coverslip, microscope and stain (saffranin).

**Procedure :** Carefully peel off a thin layer from the bulb of the onion. Initially it may be difficult to get a thin layer but after the practice it will be possible to get a thin layer which is suitable for microscope examination. For this purpose carefully separate one fleshy leaf of the onion bulb. Bend it in the middle so that it incompletely breaks. When you separate the two pieces a thin layer will come out. Immediately after removing place the onion peel in a petridish containing water.

Cut a small piece of the onion peel suitable for observation and place it in a drop of water in the centre of a clean glass slide. Add a drop of stain, put a cover glass and observe under microscope.

**Observation and conclusions.** The study shows the following structural details of the plant cell.

1. A number of rectangular cells can be seen under the field of the microscope.
2. The cells are many times longer than broad and are joined end to end and side to side to form a layer.
3. The cell wall which is the outermost layer can be clearly demarkated.
4. The cell membrane is not very clearly visible.

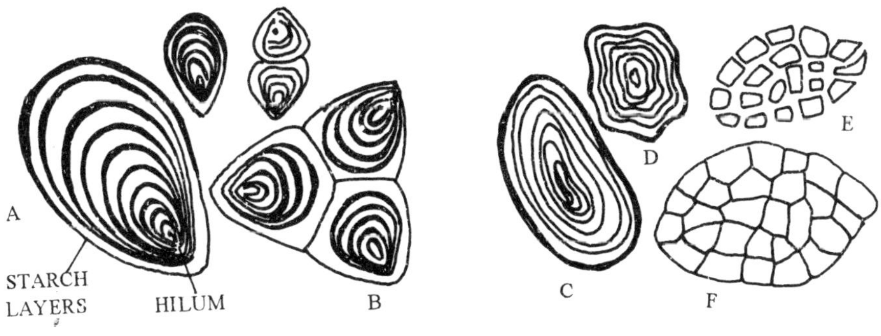


**Fig. 2.1 Plant cell**

Starch grains, A-B. Potato, C. Pea, D. Maize, E. Paddy, F. Oats.

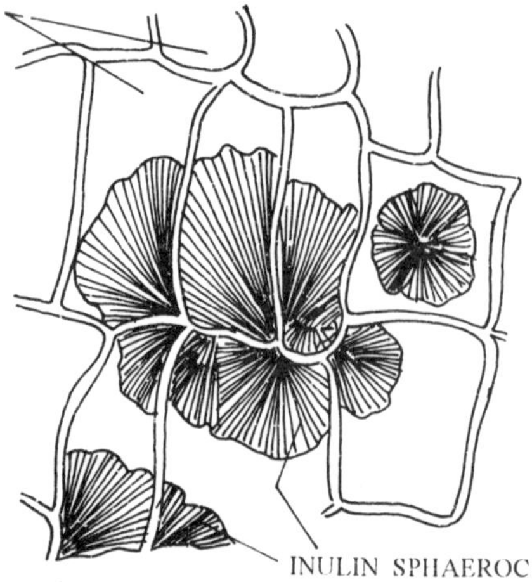


Fig. 2.2 **Plant cell**
Sphaerocrystals of inulin in the roots of Dahlia.

5.Occupying the centre of the cell is a large vacuole.

6.Cytoplasm is peripheral and occupies the space between the cell membrane and the vacuole.

7.Lying in the cytoplasm is a large vacuole which is red coloured (because it has taken saffranina stain).

8.Also distributed in the cytoplasm are seen a number of leucoplasts.

**ERGASTIC SUBSTANCES** (Non living inclusions):

Many kinds of organic and inorganic compounds occur as non living inclusions in cytoplasm. These are produced during the metabolic activities of the cell and are also known as ergastic substances. Principally, there are three types of ergastic substances, *viz., Reserve food products, Secretory products* and *Waste products.*

**1. RESERVE FOOD PRODUCTS:** These are synthesised in cytoplasm and are stored as reserve food material. These are classified into three categories *viz., Carbohydrates* (non nitrogenous products), *nitrogenous products* and *fats* and *fatty oils.*

**A. Carbohydrates:**

1. These are the principal types of food substances containing carbon, hydrogen and oxygen generally in the ration 1: 2: 1.
2. Carbohydrates may or may not be soluble in water.
3. Some of the important carbohydrates are: Starch, Inulin, Hemicellulose, Cellulose and Sugars.

**Starch:**

1. It is an insoluble carbohydrate and is found in the cells in the form of various types of grains.
2. Starch grains consist of a substances called *amylose* which turns blue on contact with iodine.
3. These are usually present in cereals and underground storage roots and stems.
4. In each starch grain there is a central spot called *hilum* on which are deposited starch layers in the form concentric rings.
5. Based on the structure, starch grains may be classified into *concentric, eccentric, simple* and *compound.*
6. In concentric grains (peas) the rings are concentric.
7. In eccentric grains the hilum is towards one side (Potato), in simple grains (wheat) there is a single hilum and in compound grains (rice) there is more than one hilum.

**Inulin:**

1. This is a carbohydrate, soluble in water but insoluble in alcohol.

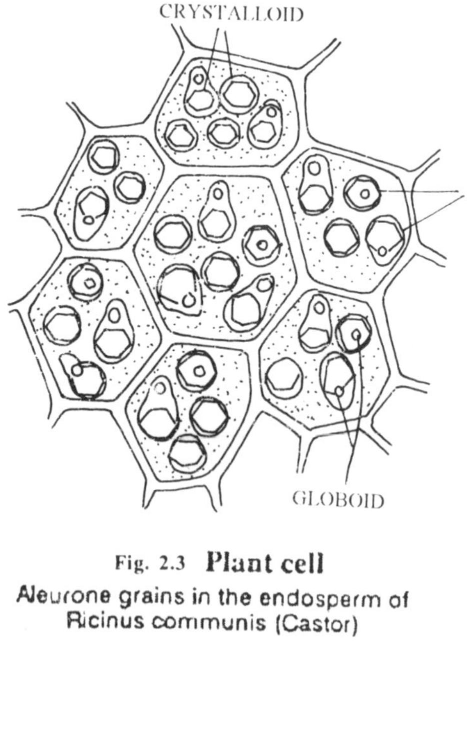


Fig. 2.3 **Plant cell**
Aleurone grains in the endosperm of Ricinus communis (Castor)

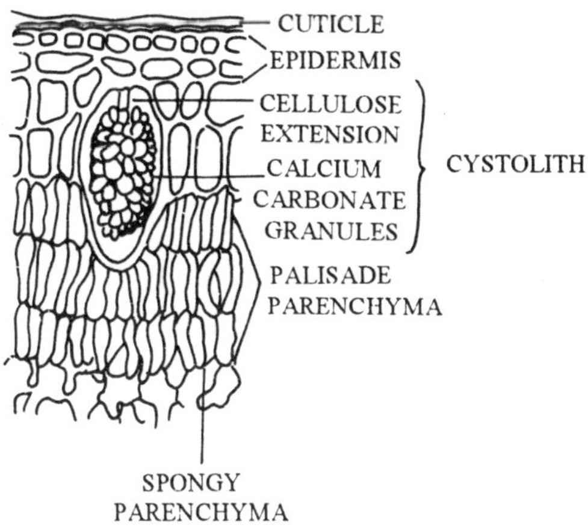


**Fig. 2.4.** Plant cell
Calcium carbonate cystolith in the leaf of Ficus elastica.

2. It is generally found in the tuberous roots of the members of the family *Asteraceae* such as *Dahlia, Helianthus, Cichorium* etc.
3. In these members inulin forms fan shaped cyrstals.
4. The inulin of *Dahlia* tubers can be precipitated by keeping them in alcohol.

**B. Nitrogenous Products:**

1. These are reserve food materials containing nitrogen.
2. There are two types of nitrogenous food materials viz., proteins and amino compounds.
3. *Proteins* are very complex macromolecules having carbon, hydrogen, oxygen and nitrogen.
4. Sometimes, in addition to these, sulphur and phosphorous are also present.
5. Proteins are stored in seeds or grains in the form of *aleurone crystals.*
6. These are present in the endosperm of castor, pericarp of maize grain etc.
7. Each aleurone grain (crystal) is a solid, oval or rounded body enclosing a small crystal called *crystalloid* and a small globular body called *globoid.*
8. The crystalloid is proteinaceous and occupies the major portion of the grain, while the globoid consists of phosphates of calcium and magnesium and occupies the rest of the grain.

**Amino Compounds** occur in the form of amino - acids and amines in the cell sap. These are more frequently present in the growing parts rather than in storage tissues.

**C. Mineral Crystals:**

1. Various types of mineral crystals are found in plant cells.
2. They may be present either in cell walls or in a cavity in the cytoplasm.
3. Majority of the crystals contain calcium carbonate or calcium oxalate.

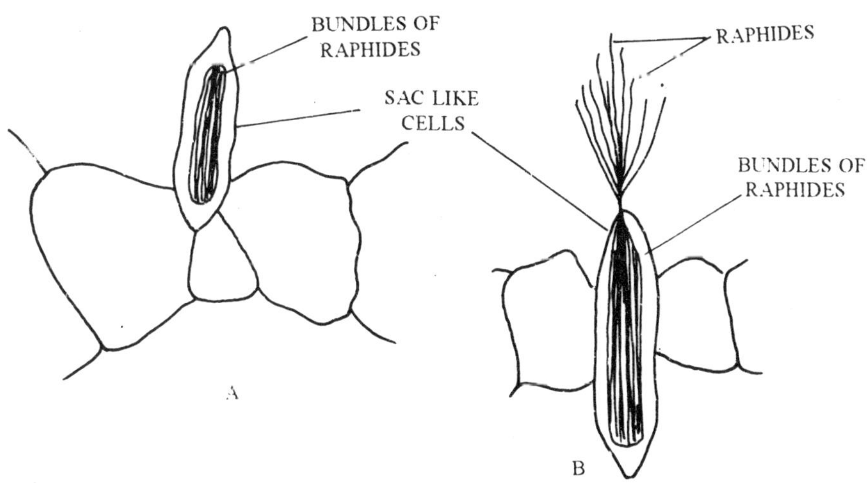


Fig. 2.5 **Plant cell**
**Raphides - A raphides in colocasia, B. ejection of a raphide.**

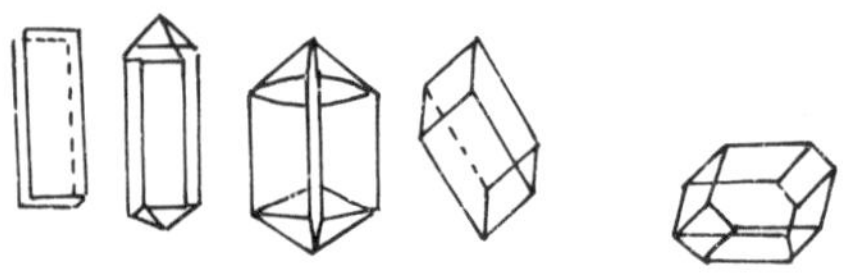

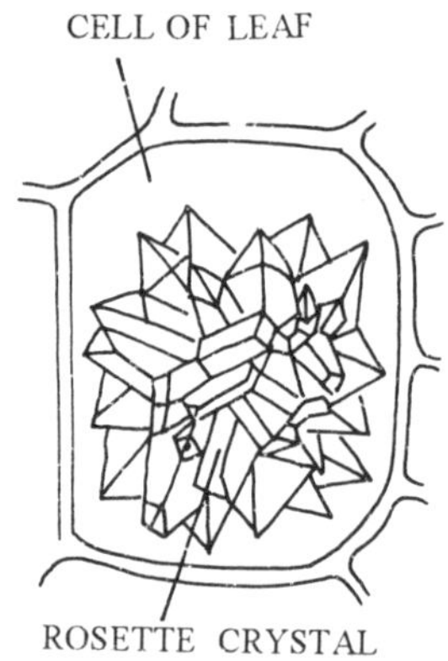


Fig. 2.6 **Plant cell**
Calcium oxalate crystals. Prismatic crystals (left) and Rosette crystal (right)

4. The following are some of the types of mineral crystals.

**1. Cystoliths:**

1. These are crystals of calcium carbonate, generally found in the leaves of various flowering plants.
2. The members of the family Moraceae, Urticaceae etc., have cystoliths.
3. Each cystolith has a peg like projection of the cell wall into the cell cavity.
4. Cystoliths consist of many small crystals grouped together like a bunch of grapes.

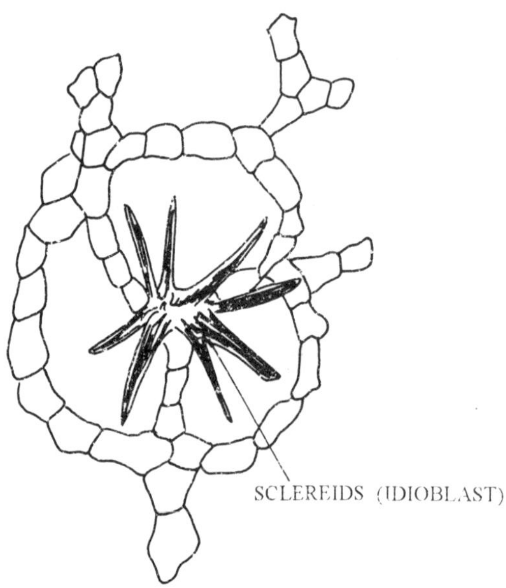


Fig. 2.7 **Plant cell**
Idioblast in a leaf cell

**2. Calcium oxalate crystals:**

1. These are found in the vacuoles and are of the following types.

**(a) Raphides:**

1. These are long needle like crystals arranged parallely in the form of stacks or bundles.
2. These are found in sacs and when the sacs are ruptured come out with a jerk.
3. Raphides are commonly seen in the leaves of *Colocasia*, *Pistia* etc.

**(b) Druses, Rosette or cluster crystals:**

1. These are compound crystals arranged in the form of a rosette.
2. These are found in the leaves of *Eucalyptus*, *Ixora* etc.
3. Crystals found in globose masses are called druses.
4. Shining crystals are called prismatic crystals.

**(c) Idioblasts:**

1. These are stellate crystals and are usually found in water plants (*Limnanthemum*, *Nymphea* etc.)

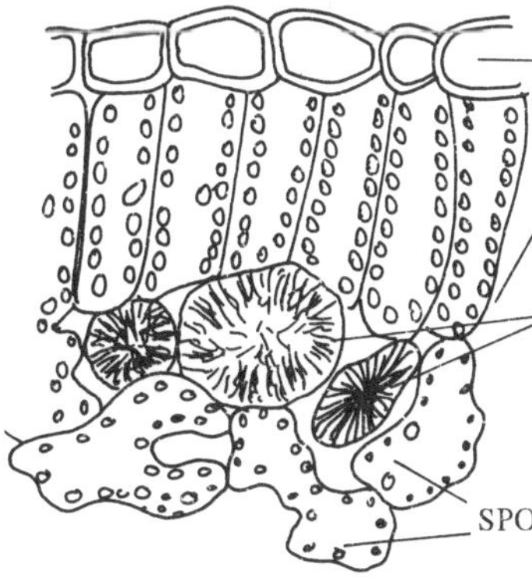


Fig. 2.8 **Plant cell**
Sand crystals of Atropa belladona.

offering mechanical support.

**(d) Sand crystals:**

1. These are found in masses packed into a cell.
2. Such crystals are seen in the roots and leaves of *Atropa belladona*.

**(e) Latex:**

1. In some plants, a milky or watery juice comes out of latex vessels or cells.
2. Latex contain starch grains proteins, oils, tannins, gums, resins, alkaloids, salts, enzymes and many poisonous substances.
3. Laticiferous cells are seen in *Carica, Euphorbia, Ficus, Hevea, Calatropis* etc.
4. In some cases the dried and processed latex is the source of natural rubber.

**(f) Essential oils:**

1. These are volatile oils and are secreted by oil glands.
2. Oil glands are found in almost all parts of the plants, specially in leaves and in the skin of fruits (*Citrus* fruits).
3. Essential oils are also found in tl

**Systematic position**
GNETOPSIDA
GNETALES
GNETACEAE
GNETUM

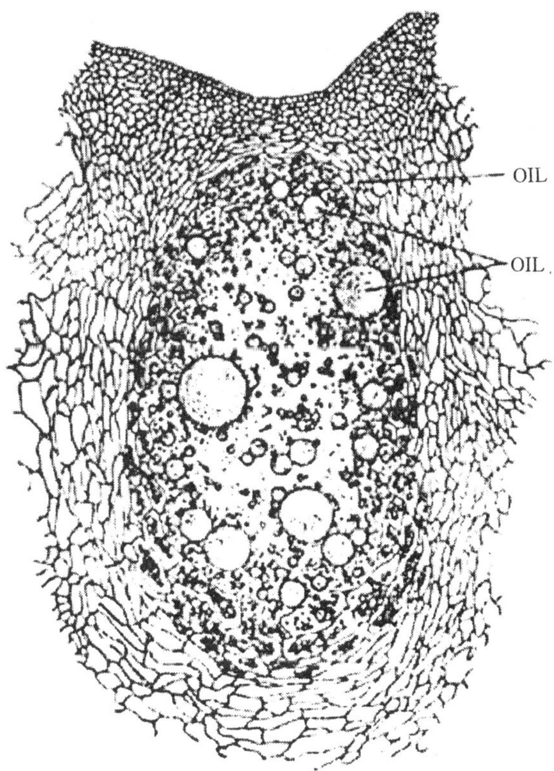


Fig. 2.9 **Plant cell**
Long section of an oil gland from Cirtus aurantiaca.

# CHAPTER 3

# TISSUES

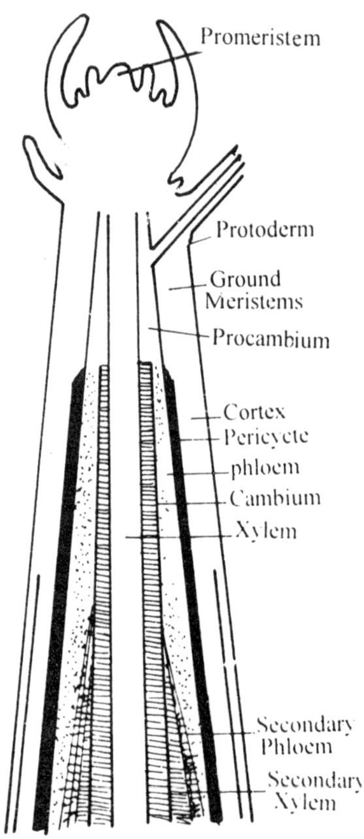


Fig. 3.1 **Meristems**
L.S. of shoot apex showing meristems based on their function

Cells are grouped together to form tissues. Generally cells which are alike in origin, form and functions constitute a tissue. In higher plants like, Gymnosperms and to a certain extent in Pteridophytes cells are of diverse types. They vary in their origin, shape, size, structure etc., Consequently the assemblage of these different types of cells form various types of tissues.

Basically tissues are classified into meristematic and permanent tissues. The basis for this classification is the stage of development of the cells. In a meristematic tissue the cells are actively dividing and have not attained maturity. The cells of a permanent tissue onthe other hand have lost the capacity of cell division and have attained their mature form.

## MERISTEMATIC TISSUES

1. The cells of a meristematic tissue are immature and are in an active state of division.
2. The cells are usually closely packed with no intercellular spaces.
3. The shape of the meristematic cells varies.
4. They may be oval, polygonal rounded or rectangular in shape.
5. The cell inclusions will be in their precursor stage.
6. The nuclei are very prominent.
7. The plane of division in the cells varies.
8. It may be periclinal (parallel to the surface) or anticlinal (Perpendicular to the surface).

**Classification of meristems:**

1. According to the location, there are three types of meristems, viz., apical, intercalary and lateral.

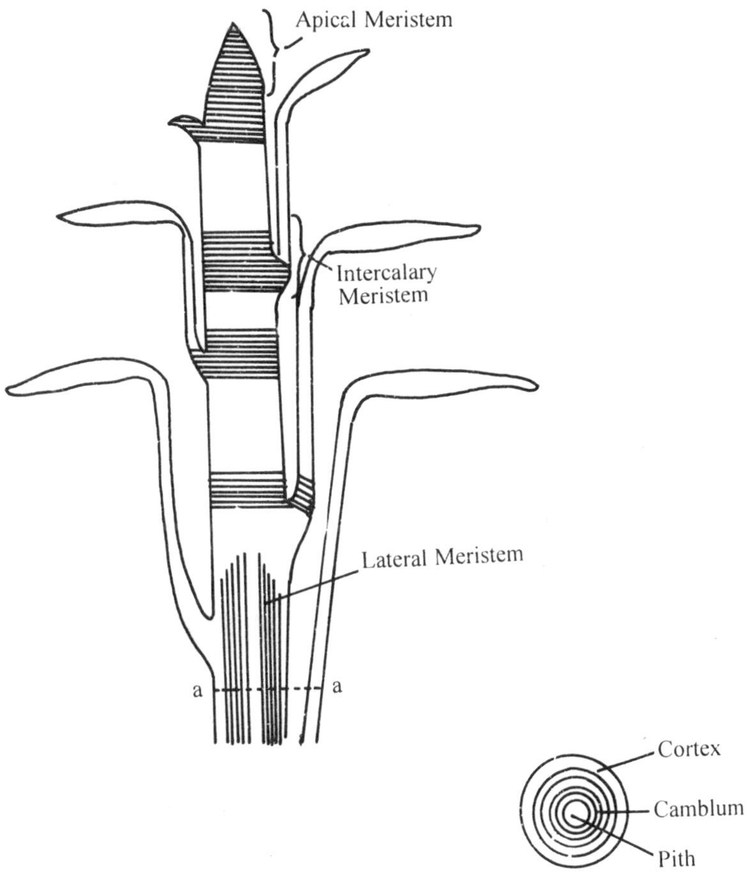


Fig. 3.2 **Meristems**

A. Diagrammatic L.S. of shoot apex showing the positions of meristems.
B.T.S. of shoot at A - A.

2. Apical meristems as the very name suggests are located at the apices of shoot, root and other organs
3. They constitute the growing points of plants.
4. The growth in length of the plant and plant organs is due to apical meristems.
5. Growth is initiated by one or more cells of the apical meristems.
6. Apical meristem consists of two zones, promeristem and meristem.
7. Promeristem consists of actively dividing cells called apical initials.
8. The meristematic zone is classified into protoderm, procambium and ground meristems.
9. Intercalary meristems represent the detached portions of the apical meristems seperated due to the growth of an organ.
10. Intercalary meristems are always internodal.
11. In the early stages of development of the plant, whole or part of the internode is meristematic.
12. Subsequently a portion of it matures, as a result, patches of meristems are found embedded in permanent tissues.
13. Lateral meristems are present along the side of the stem.
14. The cells divide in only one plane and add to the girth of the tissues.
15. Typical example of a lateral meristem is the vascular cambium.
16. Vascular cambium produces secondary Xylem and Phloem thus bringing about secondary growth.

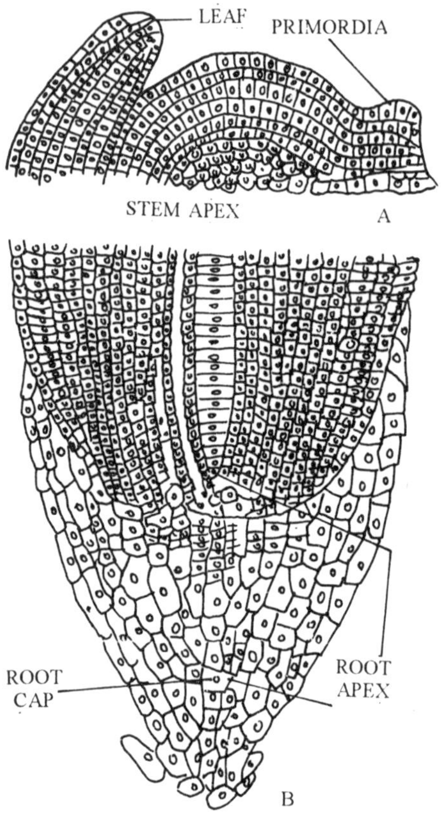


Fig. 3.3 **Meristems**
A. L.S. of shoot apex showing structure.
B. L.S. of root apex showing structure

## PERMANENT TISSUES

1. As a contrast to the meristematic tissues, the cells of permanent tissues have no capacity of division.
2. They have attained their mature form and shape.
3. The cells of the permanent tissues may be living or dead, thin or thick walled.
4. Permanent tissues may be classified into simple and complex, based on their composition.
5. A simple tissue is homogenous i.e., it is made up of only one type of cells.
6. In a complex tissue there will be at least two different types of cells.
7. But these cells coordinate and function like a single unit.
8. Generally one or more types of simple tissues combine together to form a complex tissues.

## SIMPLE PERMANENT TISSUES:

There are further classified into Parenchyma, Collenchyma and Sclerenchyma.

## PARENCHYMA:

1. Parenchyma refers to a tissue composed of living cells variable in their morphology and physiology.
2. The word parenchyma is derived from the Greek *para*, beside and *en-chein*, to pour, meaning a semi-liquid substance poured beside other tissues.

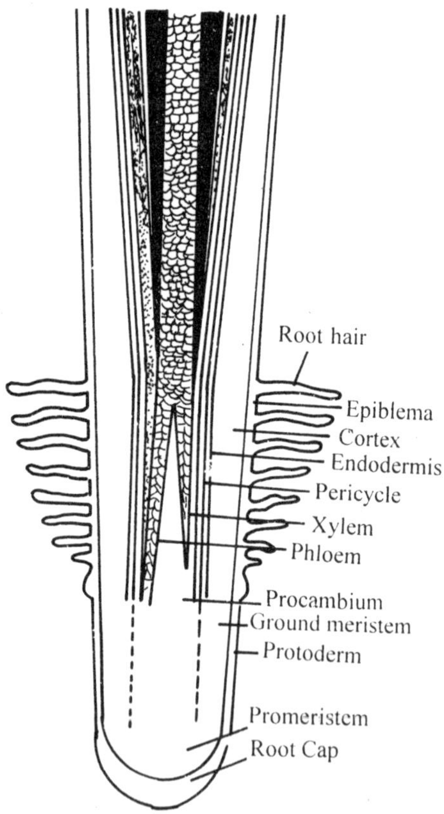


Fig. 3.4 **Meristems**
L.S. of root apex showing meristems

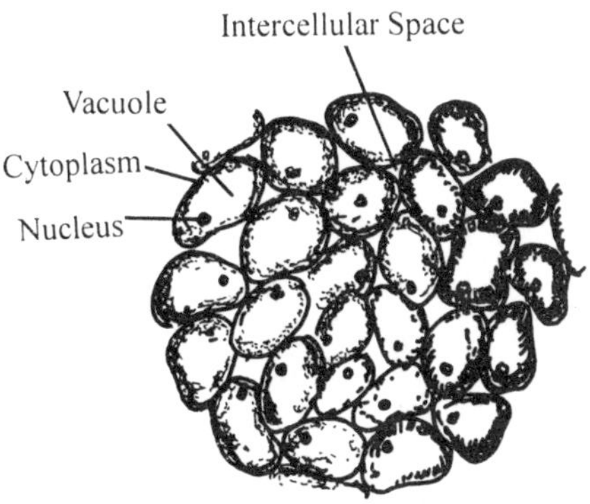


Fig. 3.5 **Permanent tissues.,**
Parenchyma

*Distribution:*

1. Parenchyma cells are present in almost all parts of the plant, like the pith and cortex of rocts and stems, mesophyll of leaves, the flesh of succulent roots, the endosperm of seeds etc.
2. As components of heterogeneous tissues parenchyma cells form vascular rays.

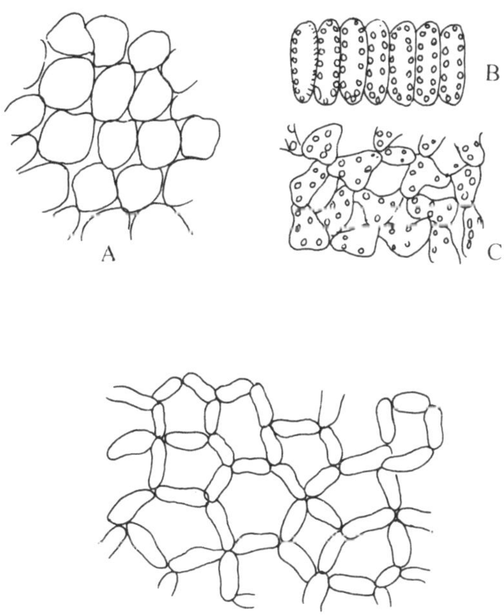


Fig. 3.6 **Permanent tissues.,**
Types of Parenchyma. A. Simple, B. Palisade, C. Spongy, D. Aerenchyma.

*Cell shape:*

1. Parenchyma cells are spherical or polygonal in shape.
2. Sometimes they may bee elongated and columnar as in the leaves.
3. Studies based on geometric models have indicated that closely packed parenchyma cells have the shape of a polyhedron with approximately fourteen faces.
4. Parenchyma cells are closely packed or arranged loosely with large number of intercellular spaces.

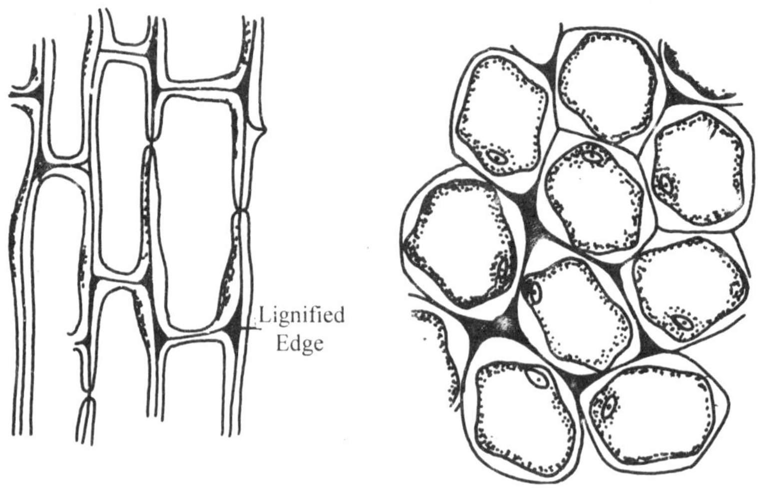


Fig. 3.7 **Permanent tissues.**

Collenchyma. A-L.S.B. - Ts.

*Cell structure:*

1. The cells have thin walls and have dense cytoplasm with a prminent nucleus.
2. Parenchyma cells in storage organs have abundant reserve food materials like proteins, fats and carbohyderates.
3. Parenchyma cells found in leaves have number of choloroplasts.
4. These help in photosynthesis.

*Cell types:*

1. Based on their structure and function parenchyma cells are classified into the following types.
2. *Simple parenchyma* which forms the ground tissue, *Chlorenchyma* which is the chief photosynthetic tissue in the plants, and *Aerenchyma* which has parenchyma cells with large intercellular spaces storing air.
3. Aerenchyma is present in hydrophytes and helps in respiration and buoyancy.
4. Chlorenchyma cells found in dorsiventral leaves are of two types viz., *Palisade parenchyma* and *spongy parenchyma*.
5. Cells of palisade parenchyma are columnar and are filled with chloroplasts.
6. Spongy parenchyma cells also have chloroplasts, but they are rich in starch grains.
7. Parenchyma cells found in Xerophytes (*Aloe, Agave* etc.) act as water storage tissues.

**COLLENCHYMA:**

1. Collenchyma is a living tissue and is composed of more or less thick primary non lignified walls.
2. Just as in parenchyma, even in collenchyma the protoplast is living.

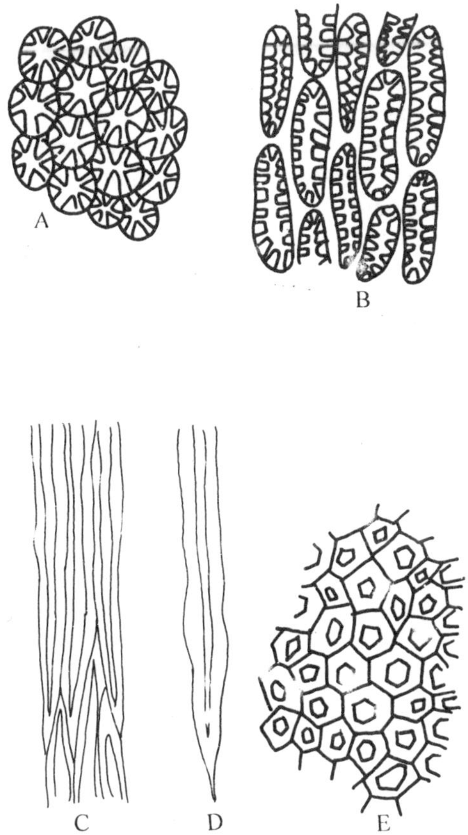


**Fig. 3.8 Permanent tissues.,**
**Sclerenchyma. A., T.S. of sclereids. B., L.S. of the same. C. and D., L.S. of fibres. E. T.S. of Fibres**

3. The word collenchyma is derived from the Greek word *colla*, meaning glue, which refers to the thick glistening wall

*Distribution:*

1. Collenchyma characterstically occurs in the pripheral position of stems, leaves, floral parts etc.
2. It may also occur in root cortex particuiarly if it is exposed to light.
3. In the stems and leaves of dicotyledons, collenchyma is hypodermal situated immediately below the epidermis or seperated from it by one or two layers of parenchyma.
4. Collenchyma cells are considerably elongated.
5. Sometimes, cells are 2 mm, long resembling fibres.
6. The shorter collenchyma cells are prismatic and appear polygonal in a transection.
7. In a long section the cells appear rectangular.

*Cell structure and function:*

1. The structure of the cell wall is the most characterstic feature of the collenchyma cells.
2. The thickenings are deposited unevenly, mainly deposited at the corners where several cells join together.
3. In long section, collenchyma cells show thin and thick wall portions dependng on the direction of the cut.
4. The walls containing mainly of cellulosic and pectic compounds and consist of much water.
5. Collenchyma cells have living protoplasts and may have chloroplastss.

6. Chloroplastss are less in number in long and narrow cells and more in cells, which approach parenchyma in form. Tannin may be present in collenchyma cells.
7. Collenchyma is primarily a mechanical tissue adapted for support of growing organs.
8. The thick walls and close packing, make collenchyma a strong tissue.
9. At the same time, the peculiarities of growth and the structure of its walls impart a considerable tensile strength with flexibility and plasticity.

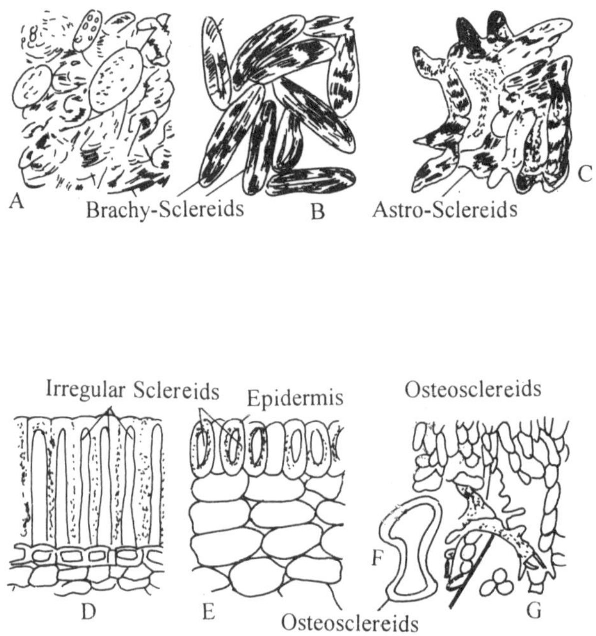


Fig. 3.9 **Permanent tissues.,**
**Types of Sclereids. A and B. Brachysclereids, C. Irregular selereids, D and E. Macrosclereids, F. Osteosclereid, G. Astrosclereid.**

## SCLERENCHYMA:

1. This tissue consists of thick walled cells, (lignified), their main function being mechanical.
2. The word is derived from the Greek *Sclerous*, hard and *Enchyma* an infusion which emphasizes the hardness of the cell walls.
3. Sclerenchyma cells show variety in form, structure, origin and development , which the different types often integrating.
4. Generally, sclerenchyma cells are classified into *fibres* and *sclereids*.
5. Fibres are considered as long cells while sclereids, are relatively short cells.

*Fibres*:

1. Fibres are long narrow, thick, lignified cells with tapering ends.
2. The wall thickenings sometimes even block the lumen of the cells.
3. The pits of fibres are simple, small round and often oblique.
4. The middle lamella is conspicuous.

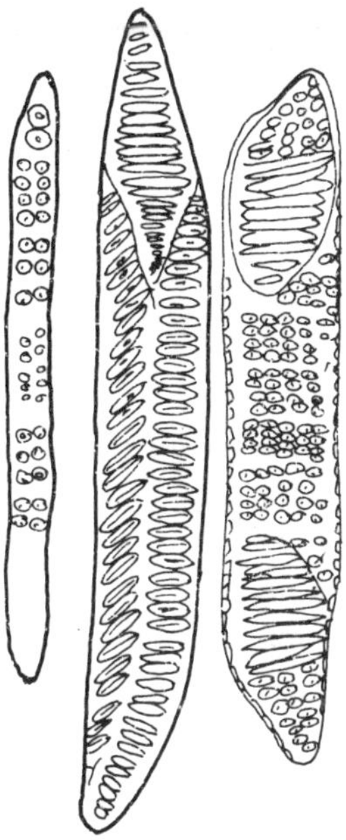

Fig. 3.10 **Permanent Tissues**
L.S. & T.S. of tracheids from *Quercus alba*

*Sclereids:*

1. These are short isodiametric cells but sometimes may be elongated also.
2. The cell walls are very thick and lignified but the cells do not have tapering ends.
3. The cells are dead and have a very narrow lumen.
4. The secondary walls show prominent pits which are simple.
5. Sometimes there is slight overarching of the secondary wall over the pit chamber.
6. The lignin deposition is uniform, but may be uneven as in the seed coats of some Leguminosae.
7. Sclereids are widely distributed in the plant body.
8. The cortex and the pith of gymnosperms and dicotyledons very often contain sclereids arranged singly or in groups.
9. Xylem and phloem may often contain sclereids besides fibres.
10. Sclereids may be present in leaves, seeds and fruits.
11. In fruits they are sometimes dispersed in the soft tissues.
12. Sclereids are classified into the following types viz., brachysclereids or stone cells (short isodiametric cells widely distributed in the flesh of fruits), macrosclereids long rod like cells found in the epidermis of leguminous seeds Osteosclereids (bone like cells) Astrosclereids (star like cells) Bizarre sclereids (Irregular sclereids), Filiform sclereids (long cells resembling fibres and trichosclereids branched, thin walled sclereids).

## COMPLEX PERMANENT TISSUES

1. The vascular tissues are the most important among the complex tissues.
2. There are mainly two components in the vascular tissue viz., xylem and phloem.

*XYLEM*:

1. This is mainly a conducting tissue of water and mineral nutrients.
2. Structurally the xylem is most complex, because it consists of various types of living and non-living cells.
3. The xylem is mainly classified into primary and secondary xylem.
4. The former differentiates in the primary plant body and is derived from the procambium, while the latter originates from a fascicular cambium during secondary growth.

*Elements of xylem:*

1. The xylem is composed of different kinds of elements viz., tracheids, fibre tracheids, vessels or tracheae, wood fibres and wood parenchyma.

(a) *Tracheids*:

1. A tracheid is an elongated tube like cell with tapering rounded oval ends.
2. The tip of the tracheid of a secondary xylem may be chisel-like.
3. The walls are thickened due to the deposition of lignin, which are of various types.

*Thickening of the cell wall*:

1. The secondary wall acquires different types of thickenings.
2. This is specially seen in cells which are to differentiate into tracheary elements i.e, vessels and tracheids.
3. In these cells, the thickening is due to the deposition of lignin.
4. The deposition of lignin is not uniform, hence we find different types of thickenings.
5. Chiefly, the following types of thickening are seen in xylem.

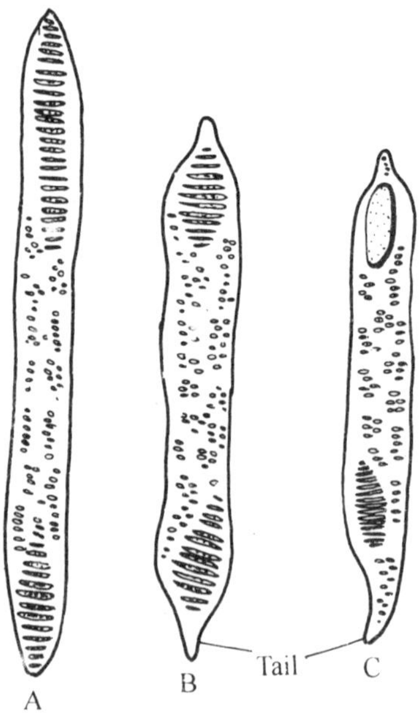


Fig. 3.1. **Permanent Tissues**

Type of vessels Simple C and Scalariform (A-B

*Annular*:

1. Lignin deposition is in the form of annuli or rings.
2. These rings are unconnected.
3. The space between the rings is unthickened.

*Spiral*:

1. Lignification occurs in the form of a spiral band.

*Scalariform* (ladder like):

1. Wall thickening is in the form of transverse rods, very much resembling the rungs of a ladder.

*Reticulate*:

1. Lignin is deposited in the form of network or reticulum.
2. The unthickened areas of the cell wall are irregular in shape.

*Pitted*:

1. Lignin deposition is almost uniform over the wall, except for small pores here and there.
2. These pores are called pits.

*Structure of pit*:

1. Pits are formed on either side of the wall.
2. In any area there will be two pits lying opposite to each other.
3. These are usually termed as "pit pairs".
4. There is a space inside each pit called pit cavity or pit chamber.
5. There is a thin membrane which seperates the pit cavities of the opposite pits.
6. It is called pit membrane.
7. The opening to a pit is called the pit aperture.
8. A pit membrane is common to both the pits.
9. There are two types of pits viz. *simple pits* and *bordered pits*.
10. A simple pit is just an opening in the wall.
11. Whereas in a bordered pit, the pit aperture is overarched by the projection of the secondary wall.
12. In a pit pair both may be simple, both may be bordered or one may be simple and the other one bordered.
13. The lost mentioned one is generally called a half bordered pit pair.
14. In a bordered pit, the pit membrane becomes thick at the centre.
15. The thickening is known as the torus.
16. Tracheids are dead cells and like sclerenchyma have neither the nucleus nor the protoplast.
17. The lumen of the cell is large.
18. The wall of the tracheids is provided with a number of bordered pits.
19. In a cross section tracheids appear polygonal or rectangular.

*(b) Fibre tracheids* :

1. During the development of fibres the wall thickening increases whereas the lumen decreases.
2. In many cells, the lumen becomes so narrow that it is altogether obliterated.
3. In such cells there is little or no conduction and they may be said to have reached the stage of true fibre.

*(c) Vessels*:

1. These are evolved from tracheids and are present mainly in Angiosperms.
2. Vessels arise through the dissolution of the end walls of tracheids.
3. Vessels are composed of rows of cells placed one above the other with the transverse walls being eliminated.
4. This facilitates the easy flow of water.

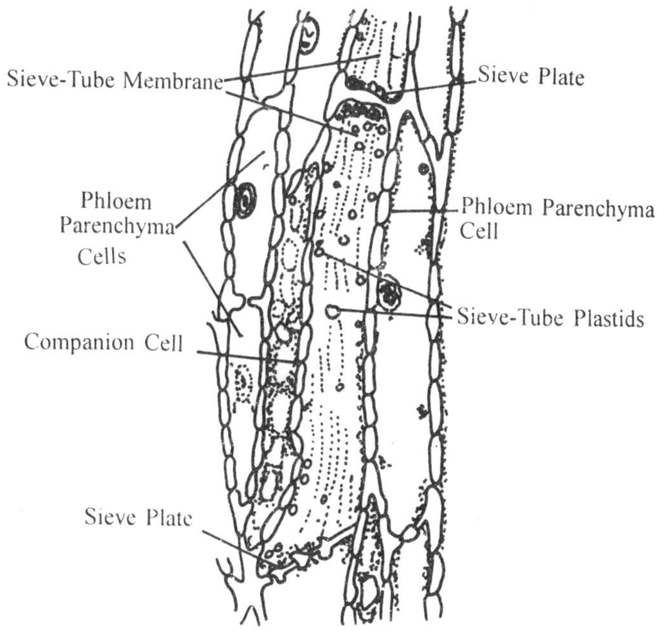


**Fig.3.12 Permanent tissues., Phloem. Long sectional View.**

5. Vessels ( tracheids also) may be of two types protoxylem and metaxylem.
6. The former matures a little early in the ontogeny and has a narrow lumen while the latter matures later and has a broad lumen.

*(d) Wood fibres*:

1. They are also called xylary fibres.
2. Sclerenchyma fibres associated with xylem are termed wood fibres.
3. They possess very thick walls and few small pits.
4. They originate from the derivatives of the cambium and have varied shapes.
5. In wood fibres, known as libriform fibres, the wall is extremely thick and lumen is almost absent.
6. Like the tracheids , wood fibres are also dead and their main function is mechanical support.

*(e) Wood Parenchyma*:

1. Parenchyma cells assicoated with xylem are called wood parenchyma.
2. These are living cells and remain so, as long as the tissue in which they lie conducts water.
3. The cells of wood parenchyma may be thin or thick walled have plenty of storage food and may help in conduction.

4. In secondary xylem wood parenchyma forms the rays which may help in radial conduction.

**PHLOEM:**

1. The phloem is the principal food conducting tissue in vascular plants.
2. Unlike xylem all the elements of phloem (except for phloem fibres) are living.
3. Like the xylem phloem is also of two types of viz., primary phloem (derived from procambium) and secondary phloem (derived from fascicular cambium).
4. The basic components of phloem are sieve elements, companion cells, phloem parenchyma and phloem fibres.
5. Occasionally sclereids, laticiferous cells and idioblasts are also found in phloem.

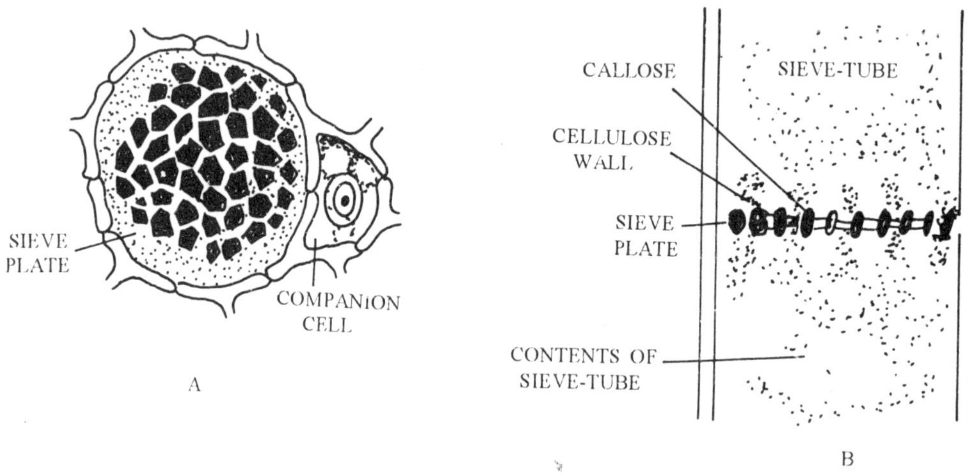


Fig. 3.13 **Permanent tissues.,**
Phloem. A. T.S. of sieve plate, B., L S. of sieve tube.

*(a) Sieve elements*:

1. What the tracheids are to xylem sieve elements are to phloem.
2. Sieve elements may be seggregated into relatively less specialized sieve cells (found in pteridophytes) and more specalized sieve tubes.
3. Seive tubes are living cells and have a thin cellulose wall.
4. In angiosperms they are enucleate, while they are nucleate in gymnosperms.
5. The protoplast of a sieve tube has a big central vacuole which is filled with viscous and albuminous proteins.
6. Leucoplasts are also reported in the vacuole.
7. The cross walls between the seieve tubes are porus and are called sieve plates as they resemble a sieve.
8. Sieve plates are usually of two types viz., simple and compound.
9. In simple sieve plates, the pores are large and the plate is horizontal or transverse in position.
10. The compound sieve plate is somewhat oblique.
11. The sieve pores are large and the plate is horizontal or transverse in position.
12. The compound sieve plate is somewhat oblique.
13. The sieve pores are very often plugged with a shining substance called *callus pad* which is usually deposited at the end of the sp.ing season.
14. Chemically callus pads are made up of a carbohydrate called *callose*.

15. At the approach of the spring, the callus dissolves.
16. Through the sieve pores, thin cytoplasmic strands come out and enter into other cells.

*(b) Companion cells*:

1. These are specialised small cells usually associated with the sieve tubes.
2. These are living, nucleate cells and are supposed to compensate the absence of nuclei in sieve tubes.
3. Companion cells are absent in gymnosperms, where albuminous clls are present.
4. The cytoplasm of cmpanion cell is dense and granular with a prominent nucleus.
5. Ontogenetically both sieve tubes and companion cells are derived from the same mother cell.

*(c) Phloem parenchyma* :

1. Parenchyma cells in association with phloem are called phloem parenchyma.
2. They are no different from other ordinary parenchyma cells.
3. In shape, the cells may be elongated, pointed, broadly cylindrical or even subspherical.
4. The cells contain oil, starch mucilage, latex etc.
5. Phloem parenchyma is absent in most of the monocotyledons.

*(d) Phloem fibres*:

1. These are also called bast fibres.

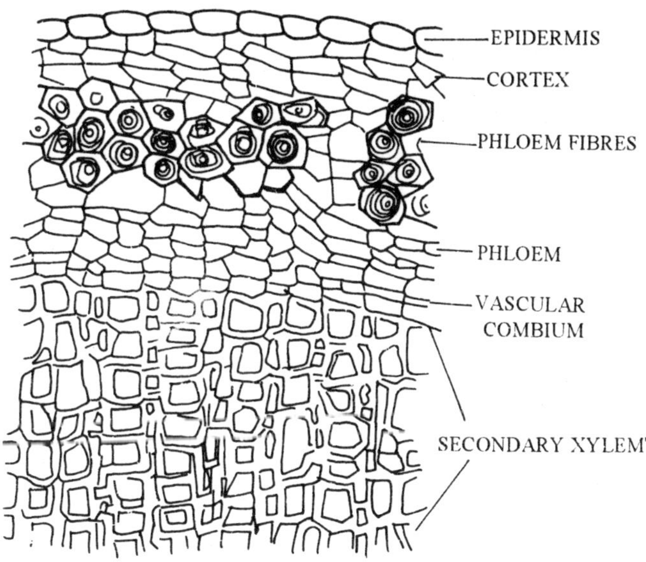


Fig. 3.14 **Permanent tissues.,**
Phloem fibres in the stem of Linum.

2. These are found both in primary and secondary phloem.
3. The walls have simple pits.
4. The bast fibres may be lignified or non lignified.

**Types of Vascular bundles**

1. A vascular bundle may be defined as grouping together of vascular elements namely xylem and phloem to form a functional unit.
2. Within a vascular bundle the two major components - xylem and phloem are arragned in various ways.
3. Depending on their relative position vascular bundles are categorized into the following types - cambium

which brings about secondary growth may or may not be present.

**Open vascular bundles :**

1. This is seen in dicotyledons.
2. Here between xylem and phloem, cambium (secondary meristem) is present.
3. As a result of this the vascular bundle is open to secondary growth.
4. There is no bundle sheath surrounding the vascular bundle.

**Closed vascular bundle :**

1. This is seen in monocotyledons.
2. Here there is no cambium in the vascular bundle.
3. Hence there is no secondary growth.
4. Further a bundle sheath surrounds the vascular bundle.
5. Based on the relative positions of xylem and phloem vascular bundles are classified into- *concentric*, *radial* and *conjoint* types.

**Concentric vascular bundle:**

1. In concentric vascular bundles xylem and phloem form concentric rings; either xylem surrounds the phloem or phloem surrounds the xylem.
2. Concentric vascular bundles are of two types-

**Amphivasal** (Leptocentric): Here phloem is in the centre surrounded by xylem.

**Amphicribal** (Hadrocentric) : Here xylem is in the centre surrounded by phloem.

Concentric bundles are always closed ie., they have no cambium.

**Radial bundles:**

1. This kind of a vascular bundle is usually seen in the roots of dicotyledons and monocotyledons.
2. Here the xylem and phloem together will not form a single bundle, instead they form alternating bundles and are present on different radii.
3. This is regarded as the most primitive type.

**Conjoint bundles :**

1. In this type xylem and phloem together form compact bundles and they are present on the same radius.
2. Conjoint bundles may be open or closed and are usually present in stems.
3. Conjoint vascular bundles of monocots are always closed (no cambium) while those of dicots are always open (cambium present).
4. Conjoint vascular bundles are of two types.

**Collateral :**

1. Here xylem and phloem lie on the same radius with xylem towards the centre and phloem towards the periphery.
2. In otherwords there is one group of phloem and one group of xylem.
3. In dicotyledons cambium is present in between xylem and phloem while it is absent in monocotyledons.
4. Collateral vascular bundles are found in stems of Angiosperms.

**Bicollateral .**

1. In these bundles there is one group of xylem surrounded by two groups of phloem - outer and inner.
2. Similarly two strips of cambia are also present.
3. Here the arrangement of various elements are as follows- inner phloem, inner cambium, xylem, outer cambium and outer phloem.
4. Bicollateral vascular bundles are usually seen in the members of the family Cucurbitaceae.

CHAPTER 4

# ANATOMY OF ROOTS

### *Root of Helianthus*

A transverse section of a young sunflower root shows the following features.

*Epidermis*:

1. This is the outermost layer.
2. This is also known as epiblema or piliferous layer and is made up of closely packed cells with no intercellular spaces.
3. Some of the epidermal cells produce unicellular extensions called root hairs.
4. These help in the absorption of nutrients from the soil.

*Cortex*:

1. Internal to the epidermis is the cortex.
2. It is made up of several layers of thin walled spherical or oval parenchyma cells with large number of intercellular spaces.
3. Some of the outer layers of the cortex may become thick walled and function as exodermis.

*Stele*:

1. Occupying the central region of the root is the vascular cylinder.
2. The outermost layer of the stele is the endodermis.
3. It is made up of a single layer of barrel shaped cells., closely packed without any intercellular spaces.
4. The endodermal cells have thickenings on their radial walls called casparian thickenings.
5. In between endodermal cells thin walled cells are present.
6. They are called the passage cells.
7. Usually they are present opposite the protoxylem.
8. Internal to the endodermis is the pericycle.
9. It is made up of a single layer of rough, thin walled parenchyma cells.
10. Surrounded by the pericycle are the vascular bundles.

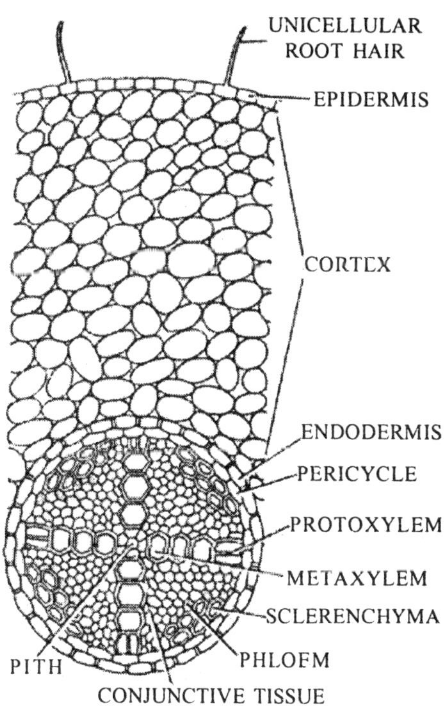


**Fig. 4.1** Anatomy of root. T.S. of dicot root. Detail of a sector.

11. They are radial (Xylem and Phloem form separate alternating bundles) with *exarch* (protoxylem pointing towards the periphery) and *tetrarch* (four protoxylem points) xylem.
12. Xylem consists of annular and sprial vessels with metaxylem being made up of reticulate and pitted vessels.
13. Occupying the central region of the stele is a narrow parenchymatous pitch.
14. Sometimes the metaxylem groups join together so that pith is obliterated.
15. Between the radial vascular bundles is present parenchyma constituting the conjuctive tissue.

***BEAN***

A transverse section of a young bean root shows the following features.

**Epidermis:**

1. This is the outermost layer.
2. This is also known as *Epiblema* or *Piliferous layer* and is made of closely packed cells with no intercellular spaces.
3. Some of the epidermal cells produce unicellular extensions called root hairs.
4. These help in the absorptionof nutrients from the soil.

**Cortex:**

1. Internal to the epidermis is the cortex.
2. It is made up of several layers of thin walled spherical or oval parenchyma cells with large number of intercellular spaces.
3. Some of the outer layers of the cortex may become thick walled and function as **exodermis**.

**Stele:**

1. Occupying the central region of the root is the vascular cylinder.
2. The outermost layer of the stele is the *endodermis*.
3. It is made up of a single layer of barrel shaped cells closely packed without any intercellular spaces.
4. The endodermal cells have thickenings on their radial walls called *Casparian thickenings*.
5. A few of the endodermal cells here and there are not thickened.
6. They are called the *passage cells*. Usually they are present opposite the protoxylem.
7. Internal to the endodermis is the *pericycle*.
8. It is made up of a single layer of rough, thin walled parenchyma cells.
9. Surrounded by the pericycle are the vascular bundles.
10. They are *radial* (xylem and phloem form separate alternating bundles), with *exarch* (protoxylem pointing towards the periphery) and *tetrarch* (Four protoxylem points) xylem.
11. Phloem consists of sieve tubes, companion cells and phloem parenchyma.
12. Protoxylem consists of annular and spiral vessels while Metaxylem is made up of reticulate and pitted vessels.
13. Occupying the central region of the stele is a narrow parenchymatous pith.

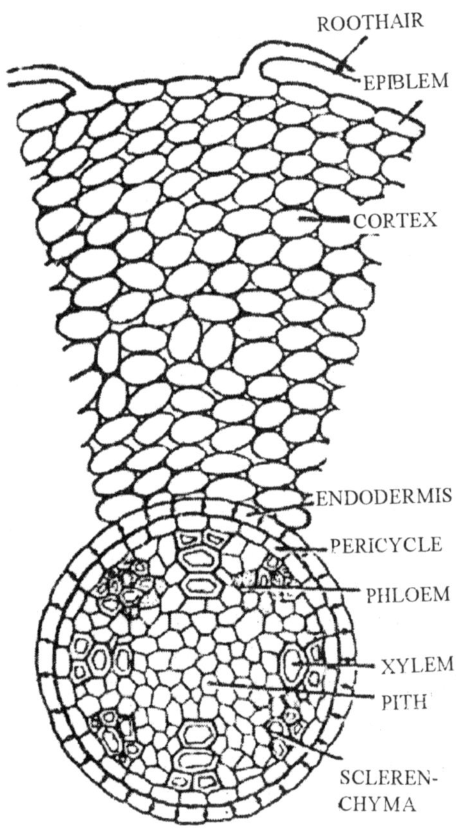


Fig.4.2 **Roots.,**
T.S. of Bean root.

14. Sometimes the metaxylem groups join together so that pi:h is obliterated.
15. Between the radial vascular bundles is present parenchyma, constituting the conjunctive tissue.

### *RANUNCULUS*

1. A T.S shows three regions - Epiblema, cortex and stele.
2. Epiblema is made up of a single layer of cells.
3. Some cells of epiblema produce unicellular root hairs.
4. Cortex is made up of many layers of rounded or polygonal parenchyma cells.
5. Some of the layers of cells immediately below the epiblema constitute exodermis.
6. The cells of the cortex is loosely arranged with intercellular spaces.
7. The cells of the cortex are rich in starch grains.
8. The stele is externally bounded by endodermis, the cells of which have casparian thickenings.
9. Next to the endodermis internally is the single layered peri cycle.
10. The vascular bundles are radial.
11. The xylem is tetrarch (four strands).
12. The xylem is exarch (protxylem points towards the perrphery).
13. Alternating with the xylem are found bundles of pholem.
14. Phloem consists of seive tubes, companion cells and phloem parenchyma.
15. Conjunctive parenchyma is present between the vascular bundles.
16. There is no pith.

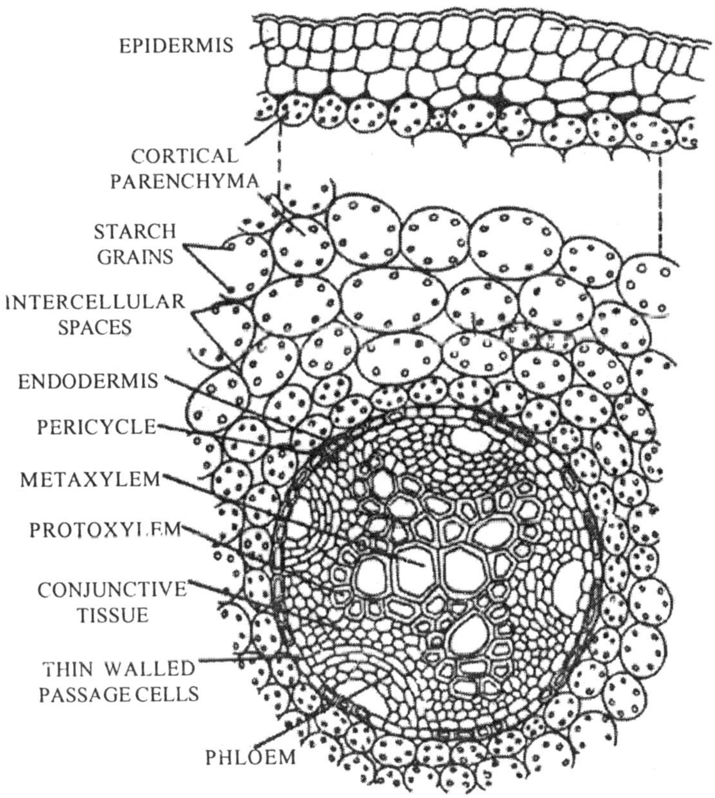


**Fig. 4.3** T.S. of Root of Ranunculus - A part enlarged

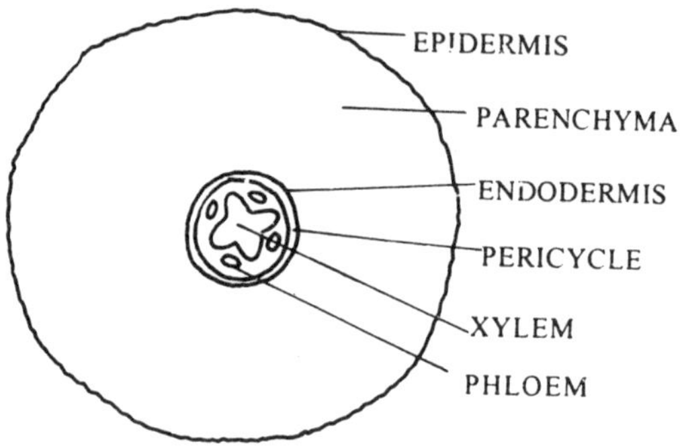


**Fig. 4.4** T.S. of Root of Ranunculus - Ground plan

## *TINOSPORA*

1. A transfer section of the root shows rhizodermis, cork, cork cambiam and vasculature.
2. The rhizodermis is the outermost layer and is made up of rectangular cells.
3. Immediately below the rhizodermis is found the cork which is made up of few layers of rectangular cells.
4. The cells of the cork are dead.
5. Internal to the cork is found the cork cambium or phellogen.
6. The cells of phellogen are actively dividing and produce cork to the exterior and secondary cortex towards the interior.
7. Endodermis and pericycle are not clearly distinguishable.

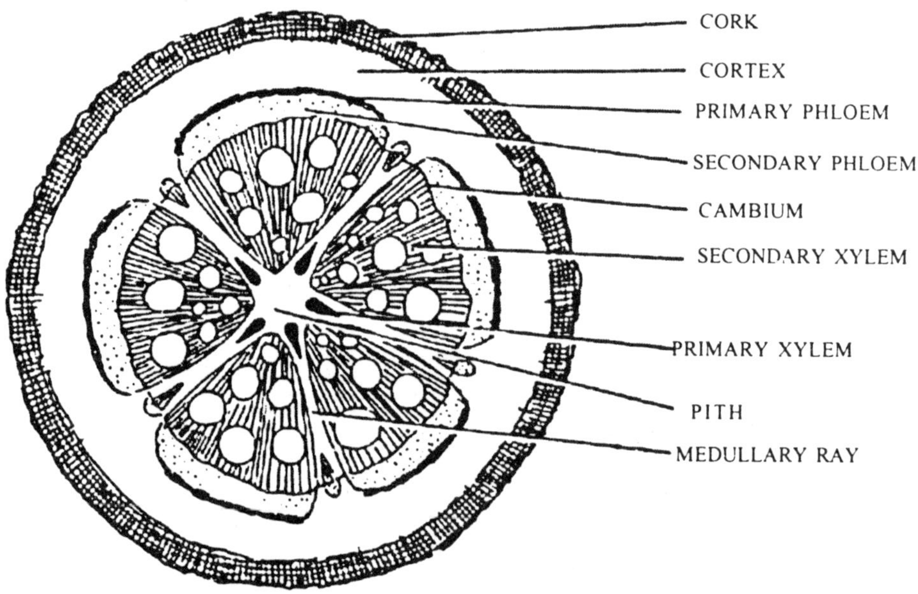


**Fig. 4.5** T.S. of Tinospora root (ground plan)

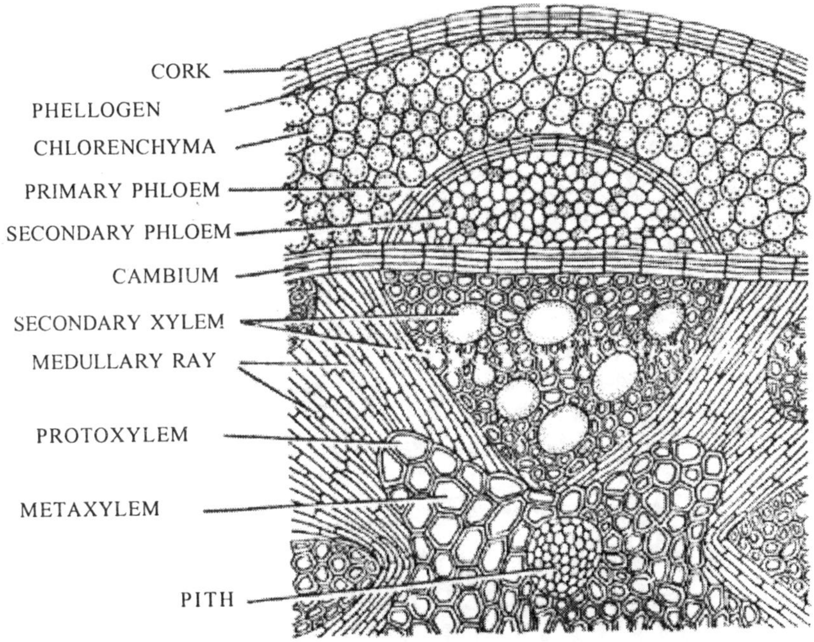


**Fig. 4.6** T.S. of Tinospora root (A sector enlarged)

8. The vascular consists of primary phloem secondary phloem and secondary xylem.
9. The primary phloem is crushed and is found next to the secondary cortex.
10. Next to the primary phloem is secondary phloem consisting of seive tubes companion cells and phloem parenchyma.
11. Next to the secondary phloem is the fascicular cambium.
12. Internal to the cambium is the secondary xylem consisting of large vessels.
13. Very close to the centre are found primary xylem elements.
14. The primary xylem is exarch and tetrarch.
15. Pith is very narrow and found in centre.
16. Between the vascular bundles are found medullary rays.

### *PROP ROOT OF BANYAN* (*Ficus bengalensis*)

1. A transverse section of the aerial root of *Ficus* in the primary structure shows an outermost epliblema followed by cortex and stele.
2. The cortex consists of parenchymatous cells which may have chloroplasts.
3. The stele is surrounded by endodermis and pericycle.
4. The vascular bundles are *radial* and six to nine in number (*polyarch*).
5. In the central region there is a well developed parenchymatous pith.
6. After the initiation of secondary growth which takes place as in other dicotyledonous roots, the following features are seen

**Cork or phellem:**

1. This replaces the epibelma and consists of 4-6 layers of suberiseed cells.
2. Certain cells of the cork contain tannin. Lenticels are also present.

**Cork cambium:**

1. This is present below the cork. It is this which produces the cork tissue.
2. The cells of the cork cambium are rectangular and actively dividing.

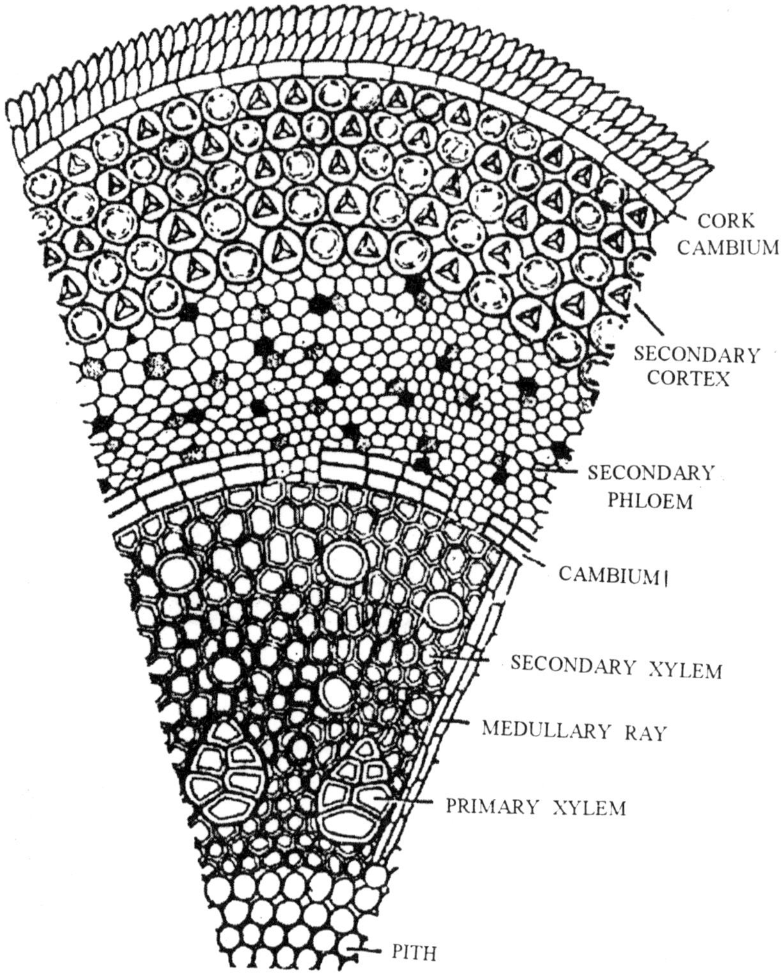


**Fig. 4.7** T.S of Banyan root (A part enlarged)

**Secondary cortex:**

1. This is also produced by the cork cambium on its inner face.
2. The cells are thin walled parenchyma, containing chloroplasts.

**Vascular tissues:**

1. After the initiation of secondary growth also, both primary and secondary vasculature can be seen.
2. The primary vascular bundles are radial with exarch, polyarch xylem.
3. Secondary vasculature forms a complete ring and is produced due to the activity of cambium.
4. In a transverse section, cambium is seen between secondary xylem and secondary phloem.
5. Fibres may be seen in secondary phloem.
6. Secondar xylem consists of vessels, tracheids and wood parenchyma.
7. In the centre there is a parenchymatous pith.
8. Medullary rays are present traversing across the secondary vascular tissues.

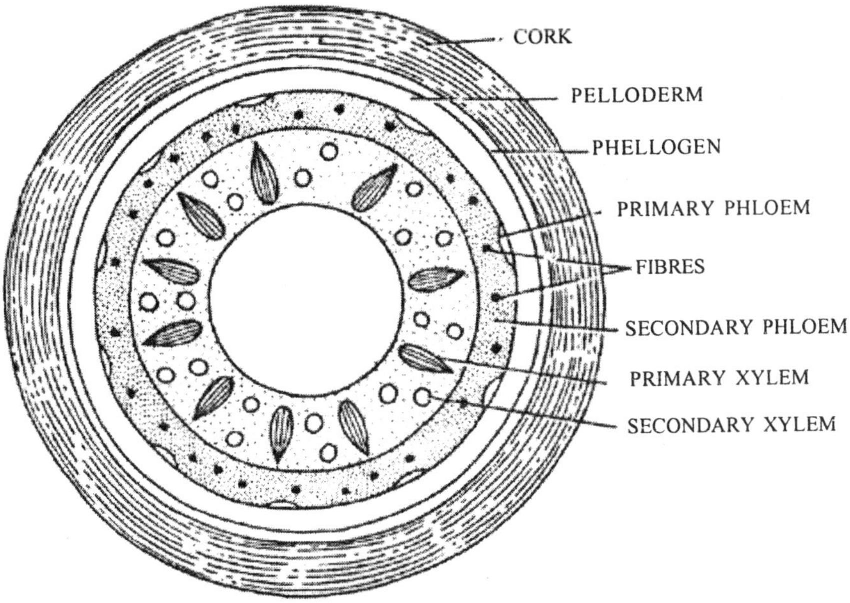


**Fig. 4.8** T.S. of Prop root of Banyan (ground plan)

### *SMILAX*

1. A transfer section shows epiblema, cortex, and stele.
2. Epibelma is single layered and it produces a number of unicellular root hairs.
3. Next to the epibelma is exodermis consisting of a single layer of thick walled cells.
4. The cortex is many layered consisting of loosely arranged thin walled parenchyma cells.
5. The cells of cortex are oval in shape and have a number of starch grains.
6. The stele is externally bounded by endodermis.
7. The cells of endodermis are barrel shaped and have casparian thickenings on their radial and tangential walls.
8. Next to the endodermis is single layered peri cycle.
9. Vascular bundles are radial and they are many in number.
10. Xylem and phloem form seperate alternative bundles.
11. The xylem is polyarch (many bundles) and exarch.
12. Phloem consists of seive tubes companion cells and phloem parenchyma.
13. Occupying the central region of the root is the pith consisting of loosely arranged parenchyma cells.

### *CANNA*

1. A transverse section shows epibelma, cortex and stele.
2. Epibelma is single layered and it produces a number of unicellular root hairs.
3. Next to the epibelma is exodermis consisting of a single layer of thick walled cells.
4. The cortex is many layered consisting of loosely arranged thin walled parenchyma cells.
5. The cells of cortex are oval in shape and have a number of starch grains.
6. The stele is externally bounded by endodermis.
7. The cells of endodermis are barrel shaped and have casparian thickenings on their radial and tangential walls.
8. Next to the endodermis is single layered peri cycle.
9. Vascular bundles are radial and they are many in number.

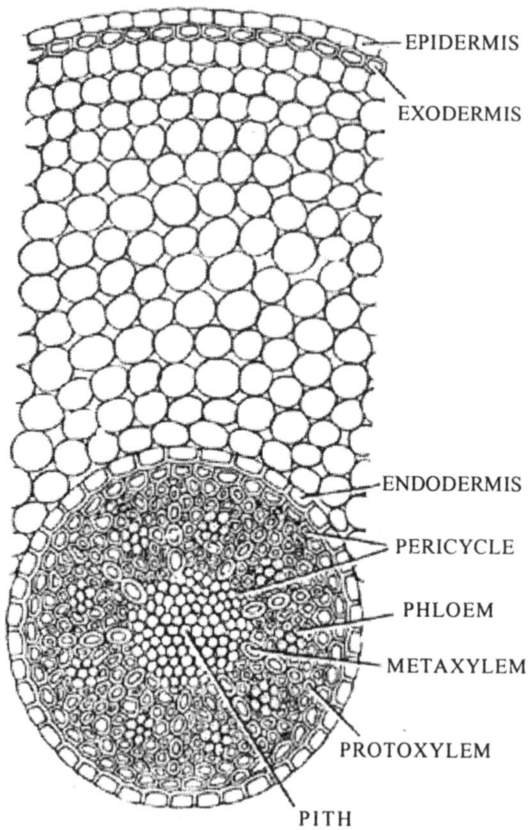


**Fig. 4.9** Anatomy of Smilax root (A sector enlarged)

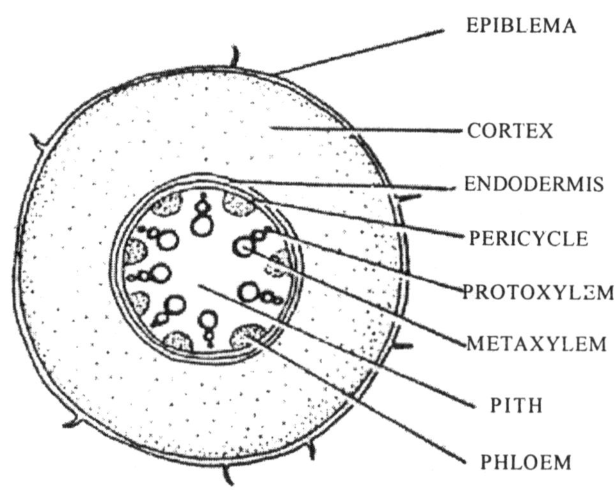


**Fig. 4.10** T.S of Cana root (diagrammatic)

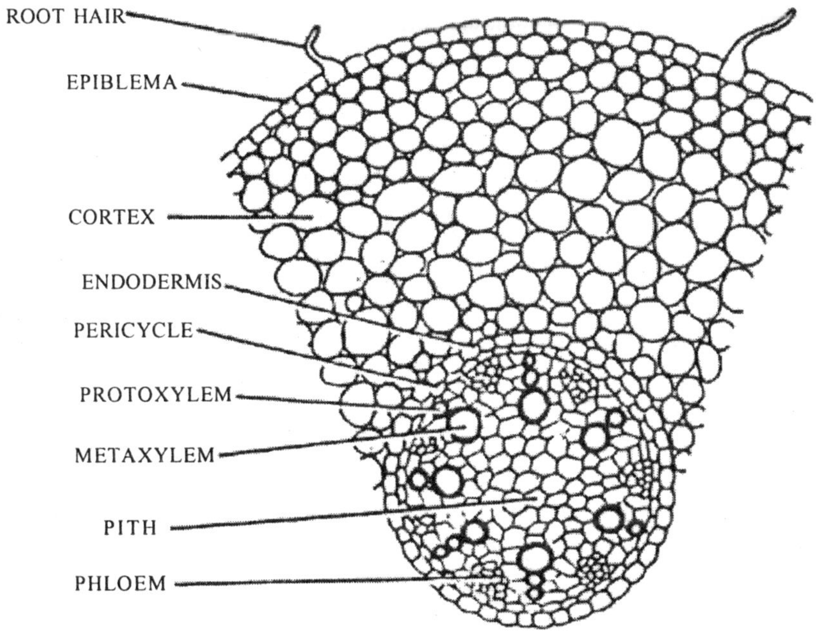


**Fig. 4.11** T.S of Cana root (A part cellular)

10. Xylem and phloem form seperate alternative bundles.
11. The xylem is polyarch (many bundles) and exarch.
12. Phloem consists of seive tubes companion cells and phloem parenchyma.

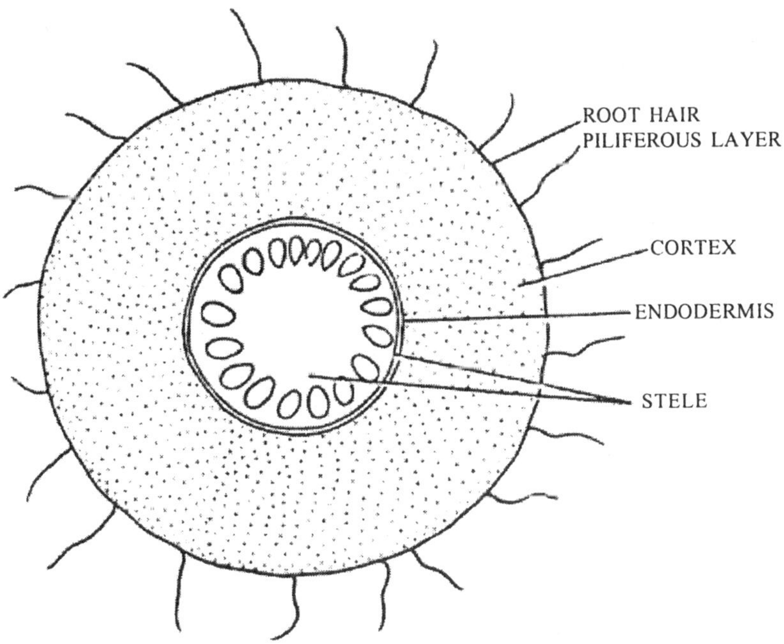


**Fig. 4.12** T.S of Maize root (Diagrammatic)

13. Occupying the central region of the root is the pith consisting of loosely arranged parenchyma cells.
14. The cells of the pith occasionally become sclerotic.

### *ROOT OF MAIZE*

1. A transverse section of a young root shows the following features.
2. Mainly three regions may be distinguished.
3. These are epiderms, cortex and stele.

*Epidermis*:

1. This is the outermost layer, and is also known as epiblema or piliferous layer.
2. The epidermis is made of a single layer of cells which are closely packed without any intercellular spaces.

*Cortex*:

1. Immediately internal to the epidermis is the cortex which is made up of many layers of loosely arranged parenchyma cells.
2. Some of the cells in the outer region of the cortex may be closely packed constituting the hypodermis.

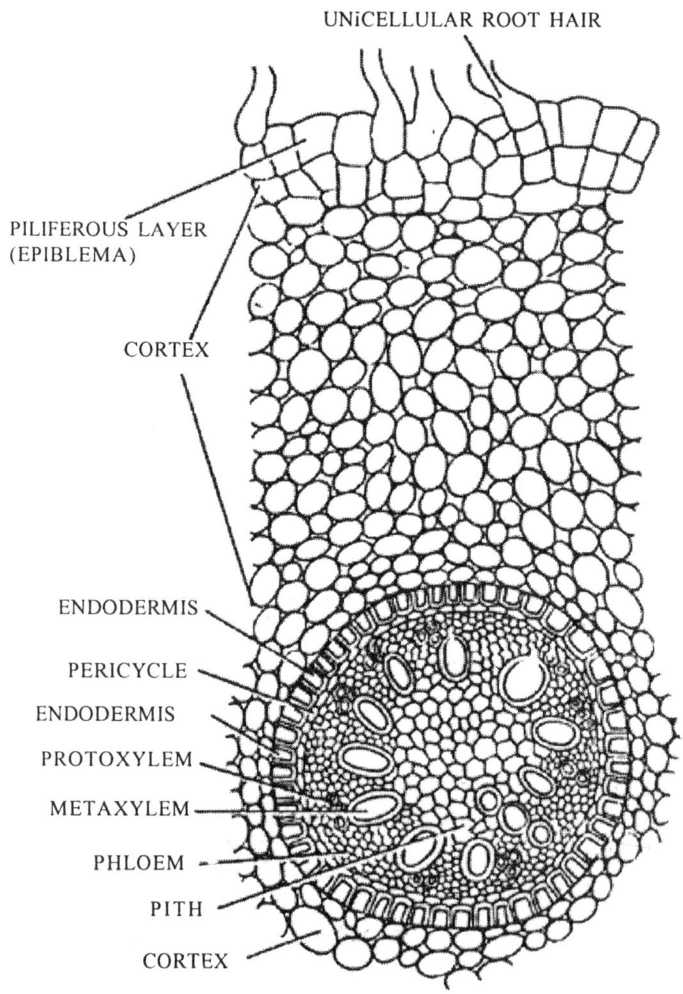


**Fig. 4.13** T.S of Maize root (Detail Structure)

*Stele*:

1. The stele is bounded by an endodermis and pericycle.
2. Endodermal cells are barrel shaped and have casparian thickenings on their radial and tangential walls.
3. Pericycle is represented by a single layer of thin walled parenchyma cells.
4. The vascular bundles are many in number (polyarch) and radial with exarch xylem.
5. In the centre of the stele is a parenchymatous pith.
6. The main points of difference between the dicotyledonous root and the monocotyledonous root are the number of xylem groups and the presence or absence of pith.

### *MUSA ROOT*

1. A transverse section of the root shows as usual epidermis, cortex and stele.
2. Epidermis or epiblema is single layered.
3. Internal to the epidermis is the cortex which is made up of many layers of loosely arranged parenchyma cells.
4. The stele is bounded by an endodermis and pericycle.
5. The vascular bundles are radial, with exarch and polyarch xylem.
6. There is a central pith.
7. In between the vascular bundles is the conjunctive tissue.
8. The conjunctive tissue as well as pith are lignified and have sclerotic cells.
9. Lying scattered in the pith are many isolated metaxylem vessels.
10. Additional phloem groups are also found scattered in the pith.

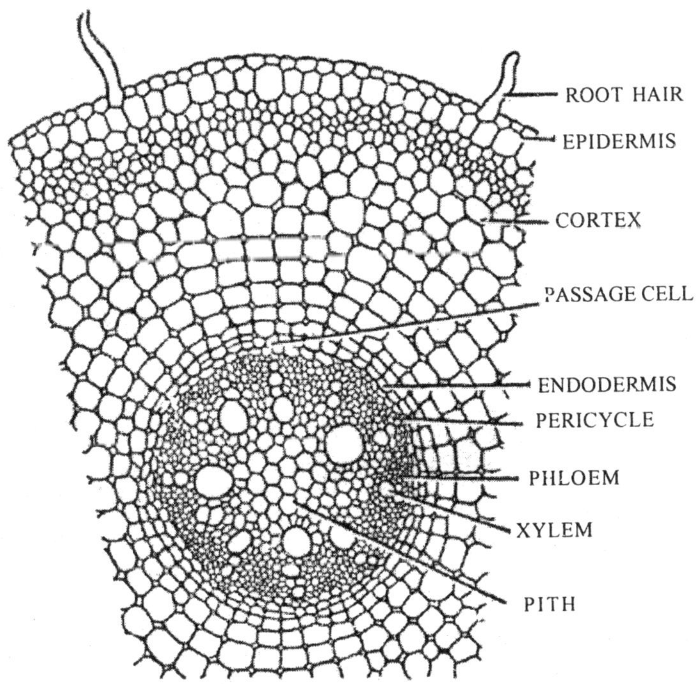


**Fig. 4.14** T.S of Musa root (A portion enlarged)

### *ORCHID ROOT* (*Vanda*)

1. The epiphytic roots of orchids have certain special features.
2. This is associated with their special function.
3. The epiphytic roots are also known as assimilatory roots isnce they help in photosynthesis.
4. A transverse section of the root of *Vanda* shows the following details:

**Velamen:**

1. This is the outermost region, and may also be calleu a multilayered epidermis.
2. The cells here are non living and elongated radially.
3. Arrangement of the cells is compact.
4. Velamen cells have the capacity to retain air and water.
5. Sometimes the cell walls develop fibrous thickenings.

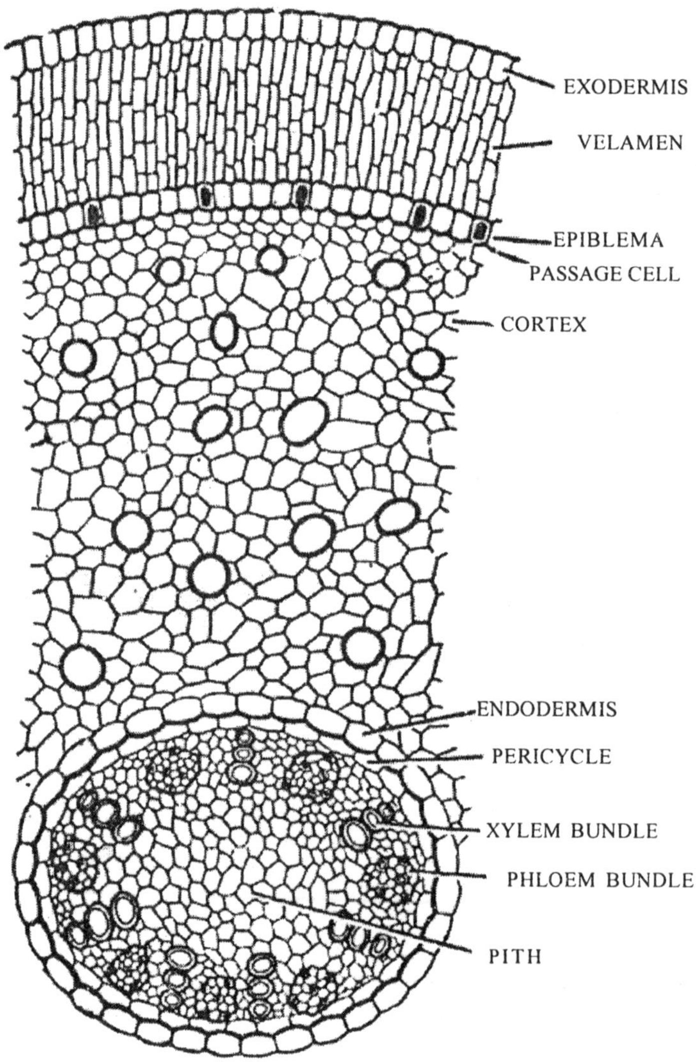


Fig4.15 T.S of Orchid root (A portion enlarged)

## Anatomy of Roots

**Exodermis:**

1. The outermost layer of the cortex is called exodermis.
2. It is made up of single layer of cells with thickened outer and lateral walls.
3. Here and there some thin walled cells are present called passage cells.

**Cortex:**

1. Internal to the exodermis is the multilayered cortex.
2. It consists of many layers of thin walled loosely arranged parenchyma cells.

**Stele:**

1. Endodermis and pericycle are present.
2. There structure is as in other roots.
3. The vascular bundles are radial with *exarch* and *polyarch* xylem.
4. There is a well developed parenchymatous pith in the centre.
5. Conjuctive parenchyma is present between the vascular bundles.

CHAPTER 5

# ANATOMY OF STEMS

## *HELIANTHUS* (Sunflower):

A transverse section of young sunflower stem shows an outmost epidermis, a middle cortex and a central stele.

*Epidermis*:

1. This is single layered with closely packed rectangular cells.
2. Above the epidermis there is a well defined cuticle.
3. Some of the epidermal cells produce multicellular hairs.
4. Occasionally stomata may be seen in the epidermis.

*Cortex*:

1. There are two regions in the cortex.

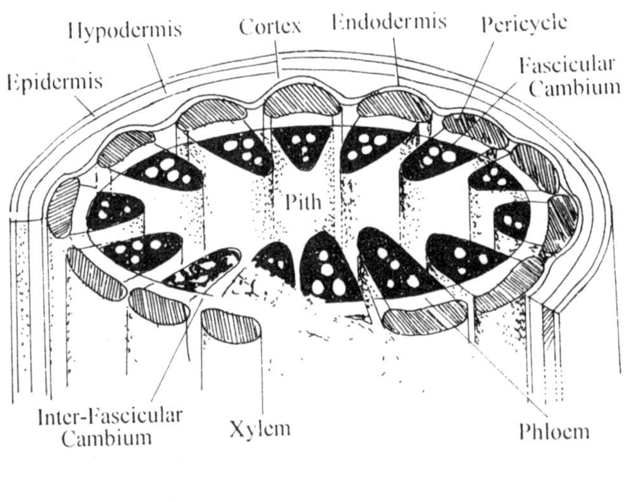


**Fig. 5.1** Stems. T.S. of Sunflower stem (Ground plan)

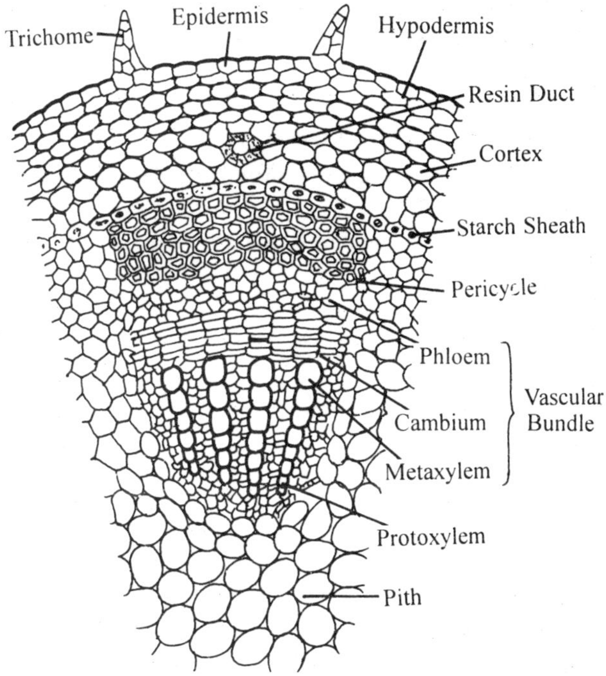


**Fig. 5.2** Stems. T.S. of Sunflower stem (A portion enlarged)

2. The outer cortex (immediately below the epidermis) consists of a few layers of collenchyma whose corner walls are thickened by the deposition of pectin and cellulose.
3. Next to the collenchyma is parenchyma constituting the inner cortex.
4. This consists of several layers of parenchyma , some of the outer layers of which are chlorophyllous.

*Stele*:

1. The outermost layer is the endodermis whose cells are rich in starch grains; hence also known as the starch sheath.
2. It has the thickening on radial walls called casparian thickenings.
3. Next to the endodermis is the pericycle.
4. It is represented by patches (cap like) of sclerenchyma present immediately above each vascular bundle.
5. The vascular bundles are many in number, arranged in the form of a ring surrounding a central pith.
6. Each vascular bundle is conjoint (xylem and phloem combining to form a single bundle), collateral (xylem and pholem on the same radius with x/p arrangement), open (open to secondary growth i.e. cabium is present) with endarch (protoxylem pointing towards the pith) xylem.
7. Phloem consists of seive tubes companion cells and phloem parenchyma .
8. In between phloem and xylem is present cambium which is said to be intrafascicular (present within a vascular bundle).
9. The cambial cells are thin walled, rectangular and actively dividing.
10. Next to the cambium, internally is the xylem, with the protoxylem facing the pith.

11. Xylem consists of vessels, trachieds, wood fibres and xylem parenchyma.
12. Between the vessels are found the wood parenchyma.
13. Between the vascular bundles are present radially elongated parenchyma cells constituting the primary medullary rays.
14. Occupying the central region of the stem is a wide parenchymatous pith.

**Stem of TRIDAX (Asteraceae)**

A transverse section shows as usual the three main regions viz., epidermis, cortex and stele.

**Epidermis:**

1. It is same as in sunflower stem, except that the epidermal hairs are much longer and two cells wide at the base.

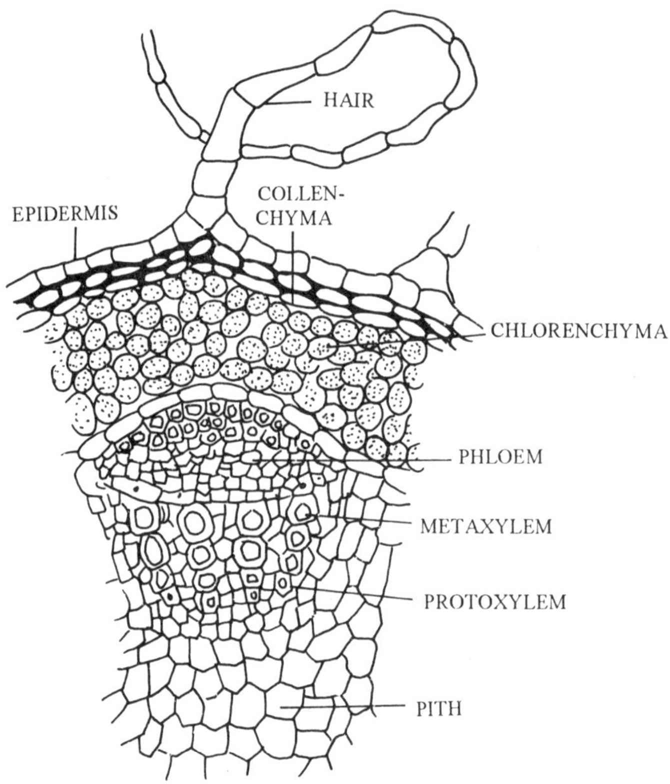


**Fig. 5.3** Stems. T.S. of Tridax stem (A portion enlarged)

**Cortex:**

1. It is also same as in sunflower stem.
2. All the cells of parenchyma are chlorophyllous.
3. Endodermis, pericycle, vascular bundles, pith and medullary rays are same as in sunflower stem.

***XANTHIUM***

1. A transverse section of a young stem shows epidermis, cortex and stele.
2. Epidermis is made up of a single layer of closely packed cells.
3. Some of the cells of epidermis produce multicellular hairs.

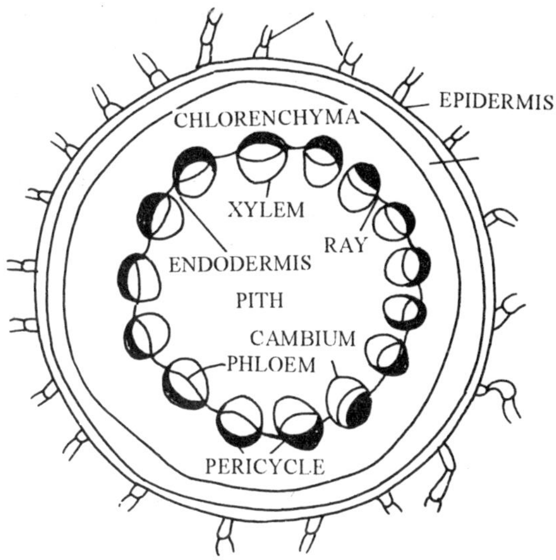


**Fig. 5.4** Stems. T.S. of Tridax stem (ground plan)

4. Some times there is deposition of cuticle above the epidermis.
5. Cortex is very broad and is made up of two zones.
6. The outer zone of the cortex consists of two to three layers of collenchyma cells.
7. The collenchyma cells are closely packed and have thickened corner walls.
8. The inner zone of the cortex consists of many layers of parenchyma cells.
9. The parenchyma cells are loosely arranged and oval in shape.
10. Bounding the stele externally is the endodermis which is not as conspicuous as in the root.
11. Next to the endodermis is the pericycle which is discontinous and found in the form of patches of sclerenchyma immediately above the vascular bundles.
12. The vascular bundles are many in number arranged in the form of a ring surrounding a central pith.
13. The vascular bundles are collateral, conjoint, and open.
14. Between xylem and phloem is present the fascicular cambium.
15. The cambial cells are narrow rectangular, and actively dividing.
16. The cambium produces secondary xylem to the interior and secondary phloem to the exterior.
17. Medullary rays are found between the vascular bundles.
18. The pith is broad and made up of loosely arranged rounded or oval parenchyma cells.

### *RICINUS*

1. A transverse section of a young stem shows epidermis, cortex and stele.
2. Epidermis is made up of a single layer of closely packed cells.
3. Some of the cells of epidermis produce multicellular hairs.
4. Some times there is deposition of cuticle above the epidermis.
5. Cortex is very broad and is made up of two zones.
6. The outer zone of the cortex consists of two to three layers of collenchyma cells.
7. The collenchyma cells are closely packed and have thickened corner walls.
8. The inner zone of the cortex consists of many layers of parenchyma cells.
9. The parenchyma cells are loosely arranged and oval in shape.

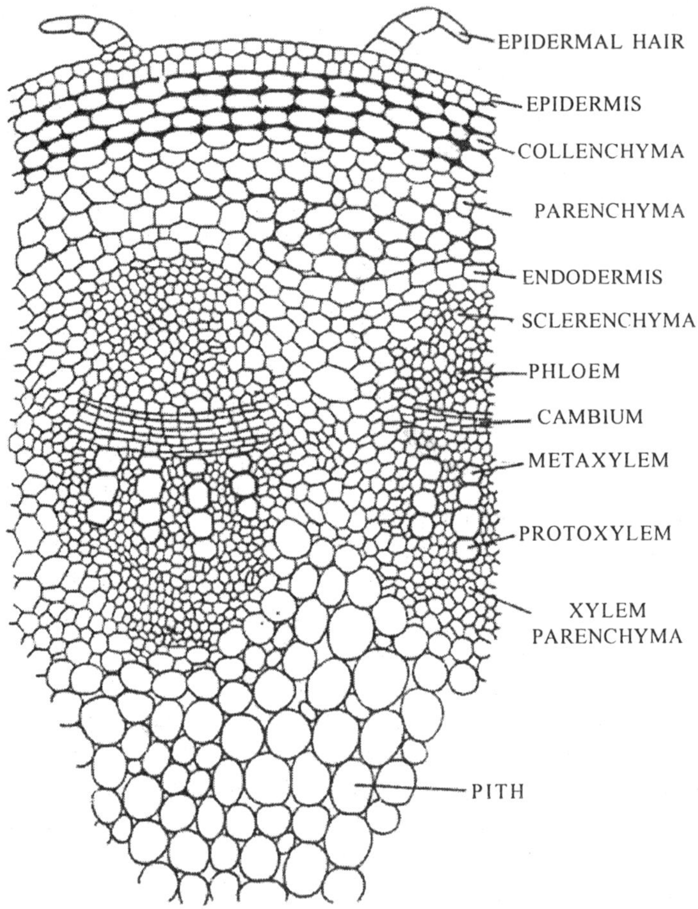


**Fig. 5.5** Stems. T.S. of Xanthium stem (Detail of a sector)

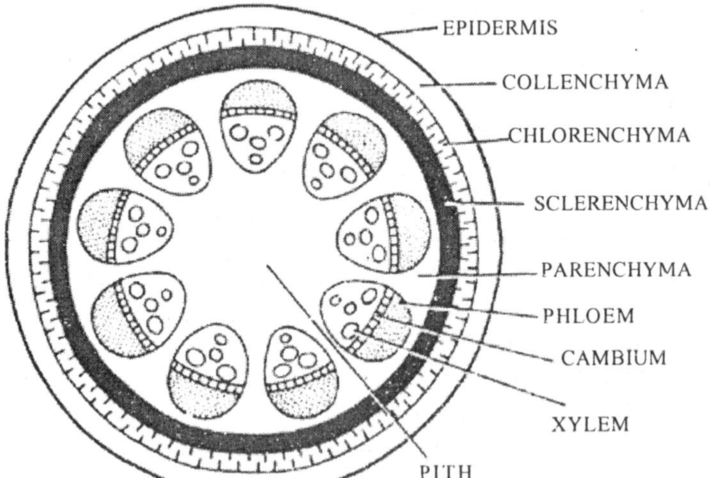


**Fig. 5.6** Stems. T.S. of Ricinus stem (Diagrammatic)

10. Bounding the stele externally is the endodermis which is not as conspicuous as in the root.
11. Next to the endodermis is the pericycle which is discontinous and found in the form of patches of sclerenchyma immediately above the vascular bundles.
12. The vascular bundles are many in number arranged in the form of a ring surrounding a central pith.
13. The vascular bundles are collateral, conjoint, and open.
14. Between xylem and phloem is present the fascicular cambium.
15. The cambial cells are narrow rectangular, and actively dividing.
16. The cambium produces secondary xylem to the interior and secondary phloem to the exterior.
17. Medullary rays are found between the vascular bundles.
18. The pith is broad and made up of loosely arranged rounded or oval parenchyma cells.

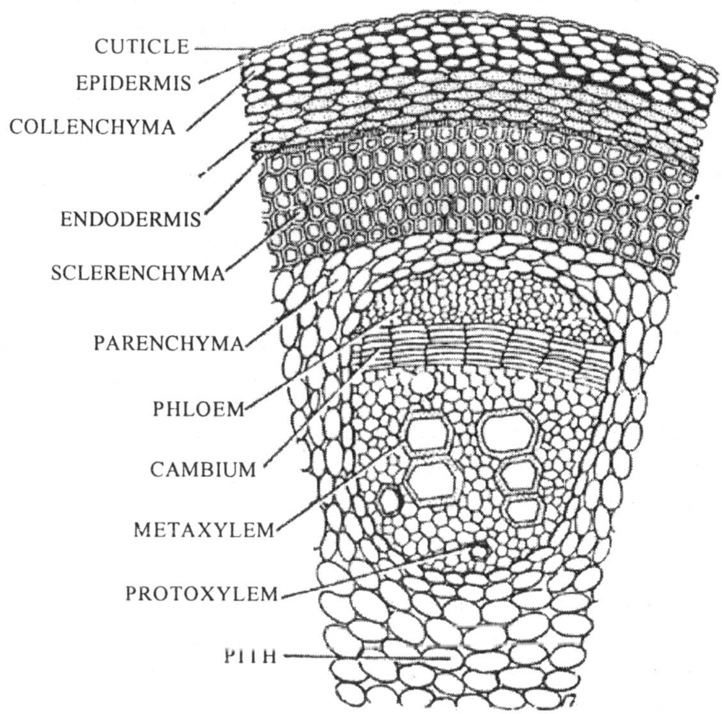


**Fig. 5.7** Stems. T.S. of Ricinus stem (A portion of sector)

### *CUCURBITA*

1. A transverse section of a young stem shows epidermis, cortex and stele.
2. Epidermis is made up of a single layer of closely packed cells.
3. Some of the cells of epidermis produce multicellular hairs.
4. Some times there is deposition of cuticle above the epidermis.
5. Cortex is very broad and is made up of two zones.
6. The outer zone of cortex consists of collenchyma cells.
7. The collenchyma is many layered near the ridges while near the forus they are either absent or very narrow.
8. The inner zone cortex consists of chlorenchyma cells consisting of chloroplasts.
9. The cells of chlorenchyma help in photo synthesis.

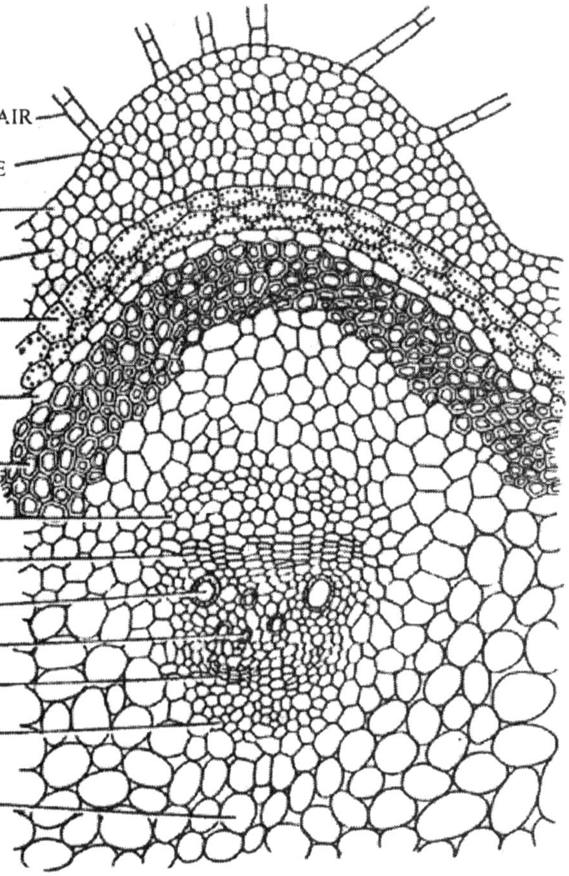


**Fig. 5.8** Stems. T.S. of Cucurbita stem (Detail of a sector)

10. Endodermis is single layered and lies immediately below the chlorenchyma.
11. Next to the endodermis is the pericycle made up of a continous patch of sclerechyma.
12. The vascular bundles are embedded in a uniform mass of parenchyma tisssues often referred to as the ground tissue.
13. The ground tissue extends from the pericycle to the central pith cavity.
14. The vascular bundles are usually ten in number and arranged in two rows.
15. Outer row corresponds to the riges and inner row corresponds to the furrows.
16. The vascular bundles are conjoint, bicollateral, and open.
17. The bicollateral vascular bundles consists of outer phloem, outer xylem, outer cambium, xylem, inner cambium, and inner phloem.
18. The xylem consists of both proto and metaxylem elements.
19. There are two strips of cambia in each vascular bundle and its activity is confined to the vascular bundles.
20. Normally the outer cambial strip in the vascular bundle is the only one which shows activity.
21. Phloem occurs in two patches, the inner one towards the pith while the outer one is towards the pericycle.
22. There is a central pith cavity.

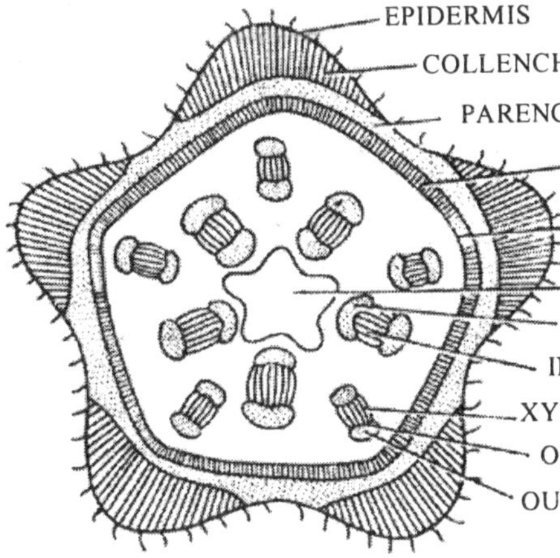


**Fig. 5.9** Stems. T.S. of Cucurbita stem (Diagrammatic)

### *TILIA*

1. The primary structure of the stem is typical to a dicotyledonous stem.
2. In the secondary structure the stem shows a rupture epidermis with cuticle deposited above it.
3. In very old stems the epidermis is completely replaced by the periderm.
4. The periderm consists of outer cork tissue, middle cork cambium, and inner secondary cortex.
5. Next to the secondary cortex is the primary cortex made up of a few layers of parenchyma cells.
6. In the secondary structure of the stem endodermis and pericycle cannot be distinguished.

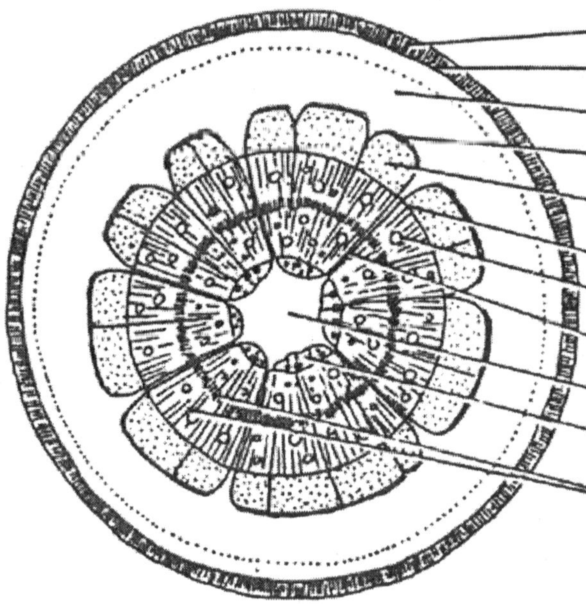


**Fig.** 0 Stems. T.S. of Tilia stem (Diagrammatic)

7. Secondary growth takes place by the activity of a fascicular cambium which forms a continous ring producing secondary vascular tissues.
8. Annular rings are produced in regular order. Each annual ring consists of spring wood and autumn wood which alternate.
9. The vessels of the spring wood are larger and the extent of spring wood is more in comparison with autumn wood.
10. Secondary phloem occurs in the form of patches seperated by triradiated medullary rays.
11. The medullary rays are of two types - triradiate occuring in phloem and narrow rays occuring in xylem.
12. The central region of the stem consists of pith made up of parenchyma cells.
13. Some cells of pith have mucilage.

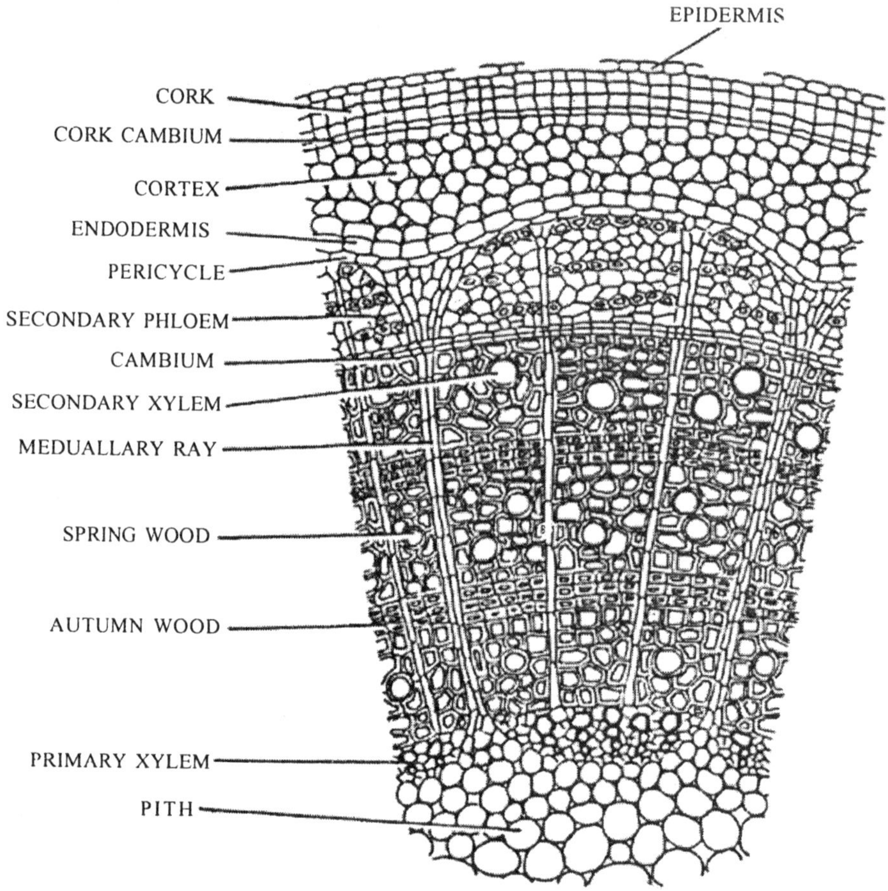


**Fig. 5.11** Stems. T.S. of Tilia stem (A part cellular)

### *STEM OF MAIZE*

1. A cross section of young stem shows an epidermis, a hypodermis and a ground tissue in which the vascular bundles are embedded .
2. There is no clear cut distinction into stele , cortex and pith. Endodermis and pericycle are absent.
3. Epidermis is made up of a single layer of compactly arranged cells.
4. It is covered with a cuticle.
5. There are no epidermal hairs.

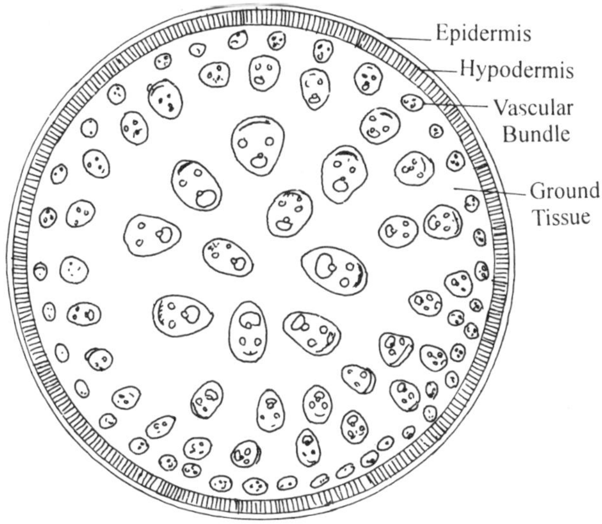


**Fig. 5.12** Stems. T.S. of Maize stem (Ground plan)

6. Hypodermis is internal to the epidermis.
7. It consists of few layers of sclerenchyma cells.
8. Ground tissue is represented by loosely arranged, thin walled parenchyma cells.
9. Embedded in this tissue are the vascular bundles.
10. Vascular bundles are scattered in the ground tissue.
11. They are crowded towards the margin but smaller in size, while towards the centre they are big in size and sparsely distributed.

*Structure of a vascular bundle*:

1. Each bundle is oval in shape, conjoint, collateral endarch and closed (cambium is absent hence no secondary growth).
2. Surrounding the bundle there is a sclernchymatous bundle sheath.
3. The phloem is above the xylem.
4. It consists of sieve tubes and companion cells.
5. There is no phloem parenchyma.
6. The outer cells of phloem (protophloem) are crushed while the inner cells (metaphloem) are healthy.
7. Xylem is reduced; it consists of four (sometimes three) vessels (two meta and two proto) arranged in the form of 'Y'.
8. Metaxylem vessels are pitted, while protoxylem vessels are annular and spiral.
9. Sometimes in between the metaxylem vessels, small, pitted tracheas are present.
10. Below the protoxylem a lysigneous cavity is present.
11. It consists water and is known as the water containing cavity.
12. Surrounding this are present cells of wood parenchyma .

## STEM OF *VANDA:*

In a cross section the following regions are seen:

**Epidermis:**

1. It is single layered.

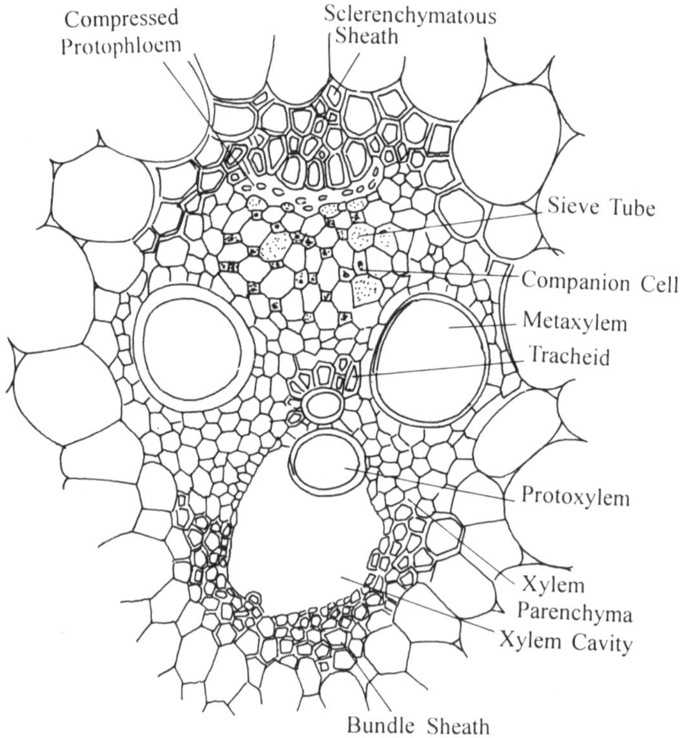


**Fig. 5.13** Stems. T.S. of Maize stem (Magnified)

2. Cuticle is deposited above the epidermis.
3. There is no hypodermis.

**Ground tissue:**

1. Below the epidermis there are one or two layers of chlorenchyma, below which is the parenchyma consisting of thin walled, loosely arranged, oval cells.
2. This zone of the ground tissue may be called the cortex.
3. The cortex is demarcated from the rest of the central parenchymatous ground tissue by means of few layers of sclerenchyma which may be called pericycle.
4. The vascular bundles are scattered in the central ground tissue.
5. A few of the bundles at the centre are larger than the rest.
6. Vascular bundles are conjoint, collateral and closed, with endarch xylem.
7. Each vascular bundle is surrounded by a sclerenchymatous bundle sheath.
8. Structure of the phloem is same as in maize stem.
9. Xylem is reduced. There is no water containing cavity.
10. Each vascular bundle has a large vessel besides tracheids.

**STEM OF *ASPARAGUS:***

In a cross section, the following regions are seen.

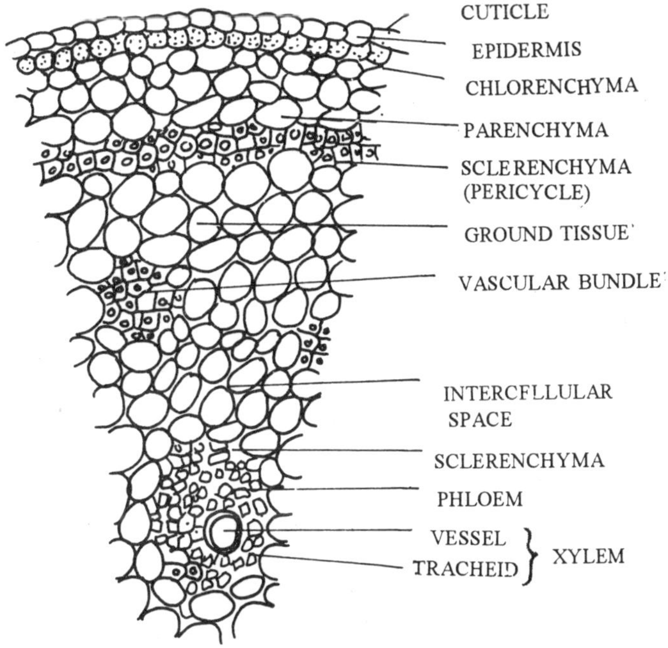


**Fig. 5.14 Stems.**
**T.S. of Vanda stem (A portion enlarged).**

**Epidermis:**

1. It is single layered and is covered by cuticle.
2. There are no epidermal hairs. There is no hypodermis.

**Ground tissue:**

1. It extends from the periphery to the centre of the axis and may be divided into four zones.
2. The outer zone which is immediately below the epidermis consists of a few layers of chlorenchyma with little intercellular spaces.

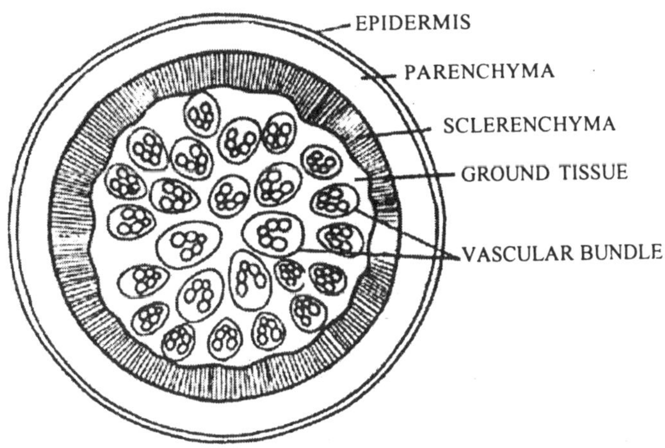


**Fig. 5.15** Stems. T.S. of Asparagus stem

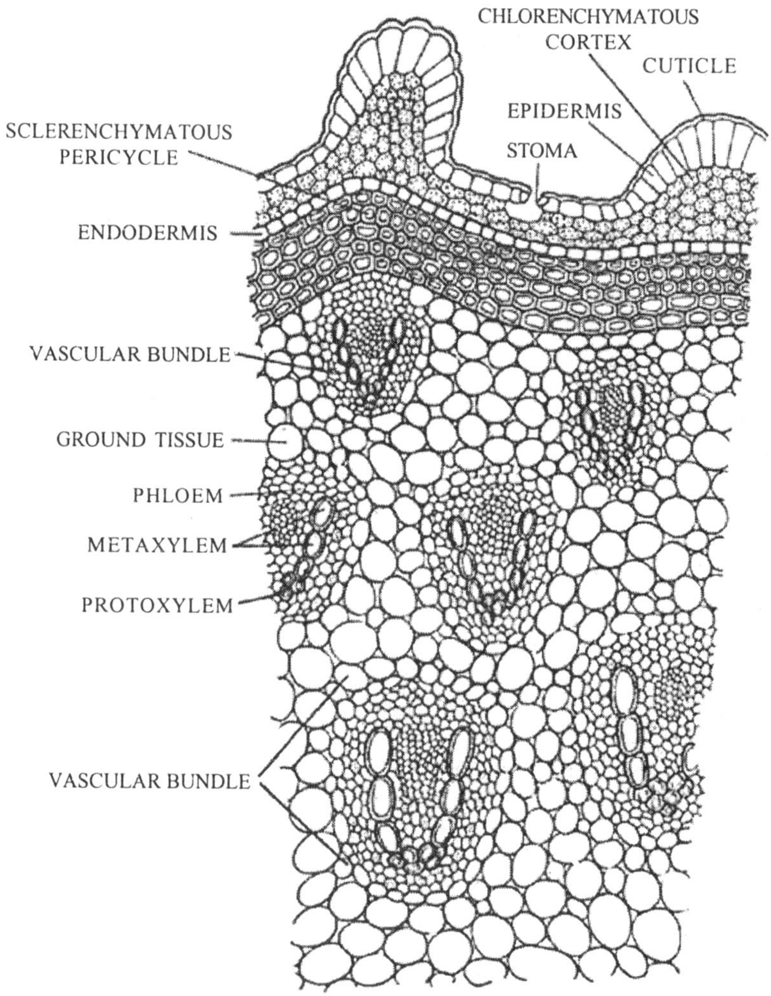


**Fig. 5.16** Stems. T.S. of Asparagus stem (vascular bundle)

3. This zone may be regarded as the cortex.
4. The second zone consists of a single layer of starch sheath which may be called endodermis.
5. Next to this endodermis is the sclerenchymatous zone.
6. The innermost zone is parenchymatous embedded in which are scattered the vascular bundles.
7. The vascular bundles are conjoint, collateral, and closed, with endarch xylem.
8. Towards the periphery of the stem they are bigger in size.
9. There is no bundle sheath.
10. The xylem strand is 'V' shaped and consists of four metaxylem vessels, a few protoxylem vessels, a few tracheids and wood parenchyma.
11. There is no water containing cavity.
12. Phloem is similar in structure to that of other monocotyledons.

### *CANNA*

1. A transverse section shows three regions namely epidermis, cortex and Stele.
2. Epidermis is the outermost layer. It is singled layered and made up of small polygonal cells.
3. The outer walls of epidermal cells are cutinized.
4. Next to the epidermis is the ground tissue.
5. Immediately below the epidermis is found the cortex.
6. The cortex consists of a few layers of parenchyma cells. These are large and polygonal.
7. Immediately below the cortex is found a single layer of chlorenchyma.
8. Next to the chlorenchyma are found patches of sclerenchyma which usually abutt the vascular bundles.
9. The entire inner region of the stem next to this consists off a uniform mass off parenchyma cells constituting the ground tissue.

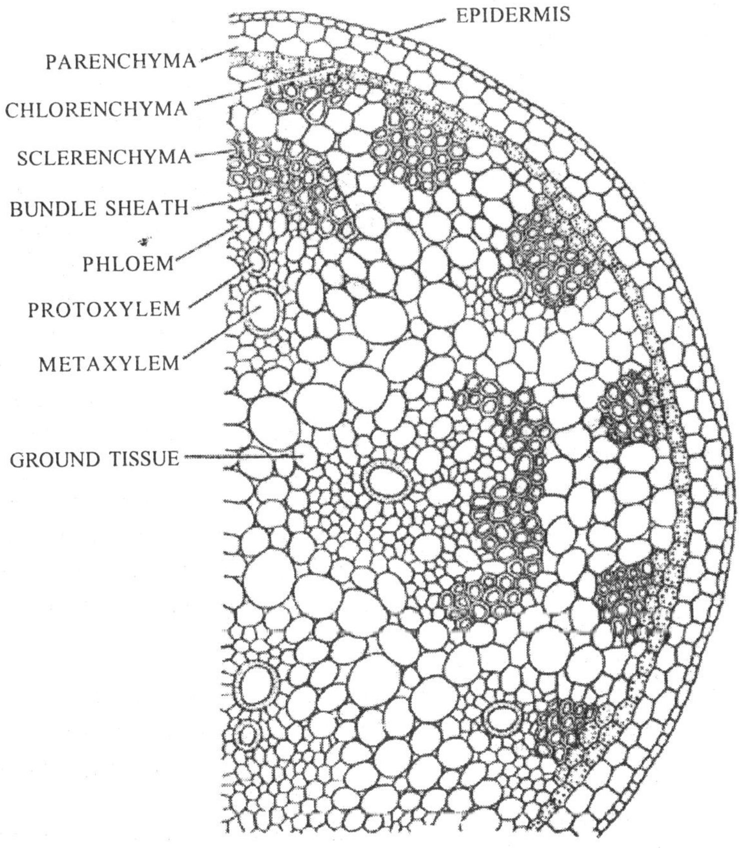


**Fig. 5.17** Stems. T.S. of Canna stem (Detail of a sector)

10. The cells of the ground tissue or loosely arranged with intercellular spaces.
11. Embedded in the mass of ground tissue are found a number of vascular bundles.
12. The vascular bundles are scattered without any arrangement.
13. The vascular bundles towards the periphery are smaller in size when compared to those in the center.
14. The vascular bundles are collateral and closed.
15. There is no cambium hence there is no secondary growth.
16. Each vascular bundle is incompletely surrounded by a bundle sheath made up of sclerenchyma.

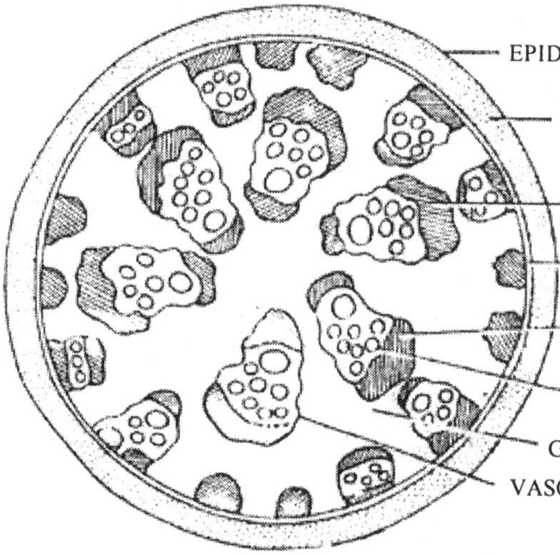


**Fig. 5.18** Stems. T.S. of Canna stem (Diagrammatic)

17. The bundle sheath actually covers the vascular bundles in the form of to caps.
18. The outer cap Of sclerenchyma is more developed than the inner one.
19. Each vascular bundle consists of xylem and phloem.
20. In the vascular bundle big xylem is found towards the inner side while the phloem is found towards the outside.
21. The xylem is made up of large spiral vessels while phloem consists of seive tubes and companion cells.

CHAPTER 6

# ANATOMY OF LEAVES

## *LEAF OF HELIANTHUS*

The leaf of sunflower shows the features of a typical dicotyledonous leaf. In a transverse section it shows three well defined regions viz. epidermis, mesophyll and veins.

**Epidermis:**

1. There are two epidermal layers covering either surfaces of the leaf.
2. The epidermis on the upper surface is called the upper epidermis and the one on the lower surface is called the lower epidermis.
3. In some cases, covering the epidermal layers will be a layer of cuticle.
4. In the leaf of sunflower the upper epidermis is continous while the continuity of the lower epidermis is interrupted by the presence of stomata.
5. There are small pores helping in gaseous exchange and the loss of water during transpiration.
6. Each stoma is covered by two guard cells which regulate the opening and closing of the stomatal aperture.

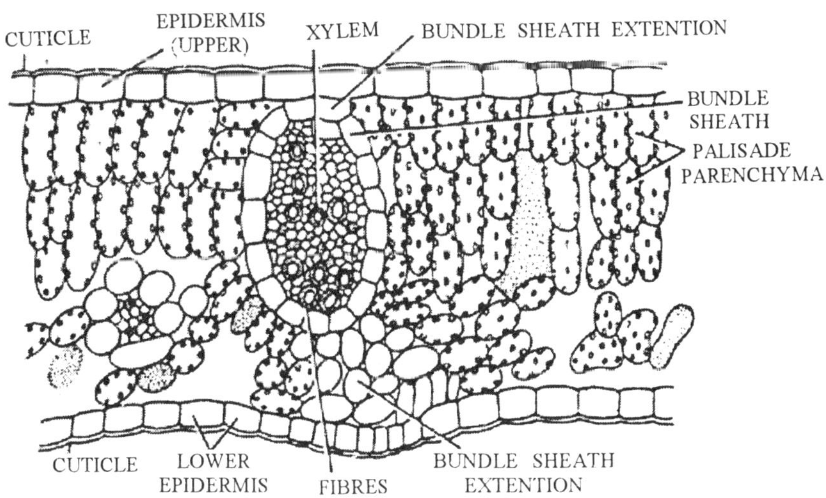


**Fig. 6.1** Anatomy of Sunflower leaf (A portion enlarged)

7. The guard cells are crescent shaped with thick inner walls.
8. Internally each stoma leads into an air space.

**Mesophyll:**

1. In between the two epidermis layers is the mesophyll.
2. It is differentiated and has well developed palisade tissue below the upper epidermis.
3. Above the lower epidermis and below the palisade tissue is the spongy parenchyma which has loosely arranged cells with large number of air chambers.
4. Palisade tissue is chlorophyllous and it is the chief photosyntietic tissue of the plant.
5. There are a few chloroplasts in the spongy parenchyma cells also but their main function is storage.
6. There are rich in starch grains.

**Vein:**

1. The veins are arranged in the form of a network and in a cross section they show the vascular bundles.
2. Each vein representing a vascular bundle is surrounded by a parenchymatous envelope called the bundle sheath.
3. Within the bundle sheath are the vascular elements viz, the xylem and the phloem.
4. Unlike in stems the xylem is above and the phloem is below.

## LEAF OF *DALBERGIA:*

The leaf of *Dalbergia* may be studied as an example of a typica dicot leaf. A cross section of the leaf shows the following features.

**Epidermis:**

1. There are two epidermal layers viz., upper and lower.
2. Both the layers are uniseriate and composed of barrel shaped cells with thick outer walls.
3. There is a well developed cuticle above the upper epidermis.
4. The leaves are hypostomatic (stomata found only on the lower surface).
5. The stomata internally lead to air cavities.
6. In between the two epidermal layers is the mesophyll.
7. It is differentiated and has a well developed palisade tissue below the upper epidermis.
8. Above the lower epidermis and below the palisade tissue is the spongy parenchyma which has loosely arranged cells with large number of air chambers.

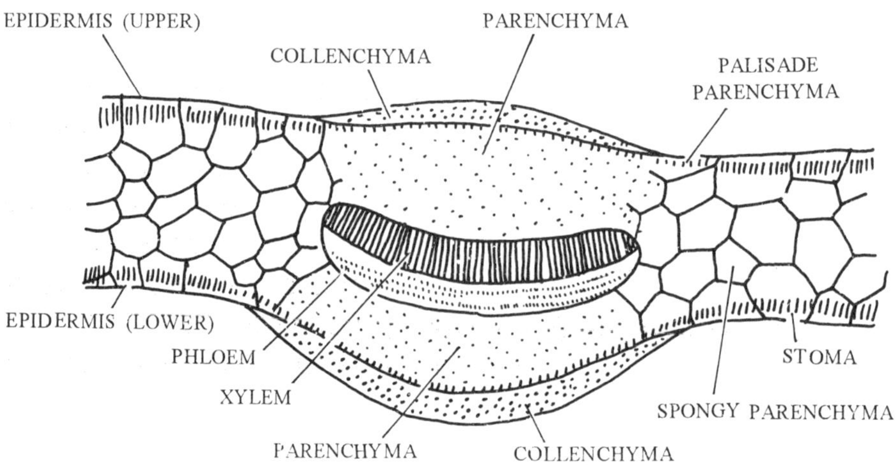


**Fig. 6.2** Anatomy of Dalbergia leaf (Ground plan)

9. Palisade tissue is chlorophyllous and it is the chief photosynthetic tissue of the plant.
10. There are a few chloroplasts in the spongy parenchyma cells also but their main function is storage.
11. They are rich in starch grains.
12. Near the midrib region, internal to the epidermis is collenchyma.
13. Collenchyma cells are more towards the lower surface.
14. Immediately surrounding the vein is a bundle sheath which is parenchymatous.
15. It is very distinct in *Dalbergia* leaf.
16. The vascular bundle is *cresent* shaped and shows x/p arrangement.

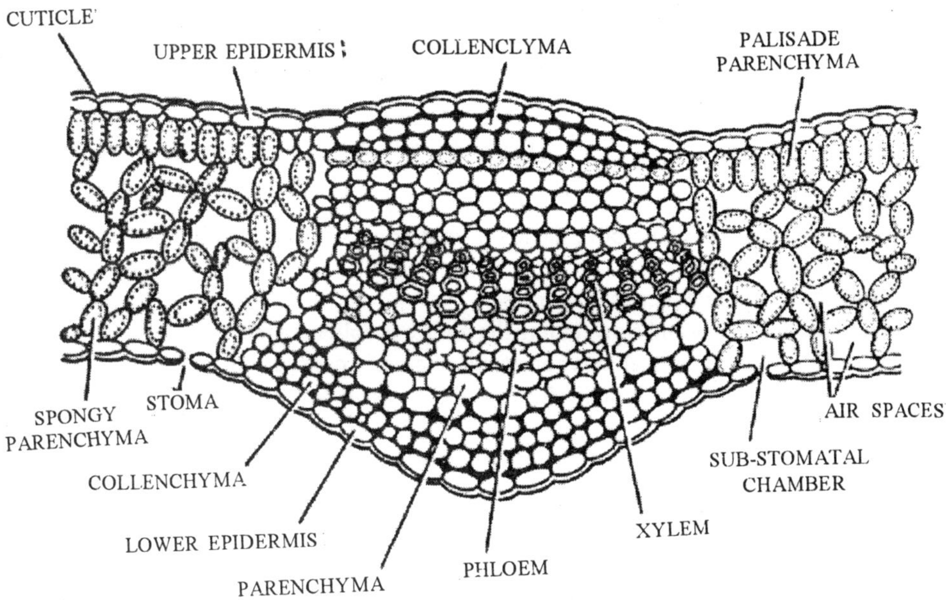


**Fig. 6.3** Anatomy of Dalbergia leaf (A portion enlarged)

## LEAF OF *NERIUM* (Oleander)

Nerium belongs to the family *Apocynaceae* of dicotyledons. It shows many xerophytic adaptations. A cross section of the leaf shows the following:

**Epidermis:**

1. The leaf is dorsiventral and shows two epidermal regions viz., upper and lower.
2. The upper epidermis is multilayered and it consists of two to three layers of barrel shaped cells arranged compactly.
3. The outermost epidermal layer is covered with a thick cuticle.
4. The structure of the lower epidermis differs from that of the upper in that it is single layered.
5. There is the deposition of a cuticle even above the lower epidermis.
6. At certain places the epidermal cells incurve forming depressions of pits.
7. The stomata are found in these depressions (hypostomatic).
8. The pits are protected with a number of *trichomes* (hairs) growing from the sides.
9. The presence of the trichomes retards the direct entry of the light and hence checks transpiration.

**Mesophyll:**

1. It is differentiated into palisade and spongy parenchyma.
2. Usually two or three layers of palisade tissue is present below the upper epidermis.

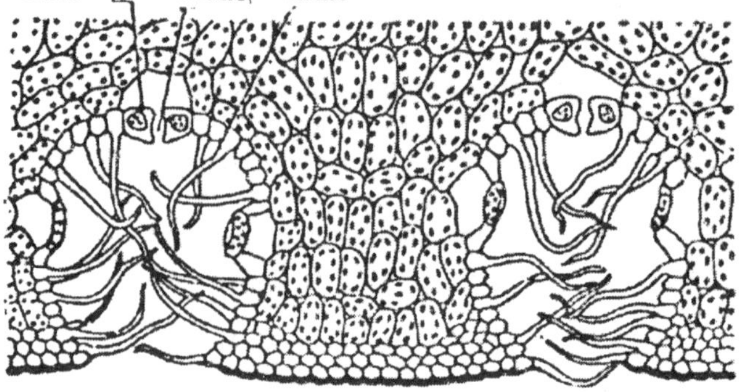


Fig. 6.4 **Leaves.,**
T.S. of Nerium leaf.

3. The spongy parenchyma consists of loosely arranged cells.
4. Sometimes, one layer of cells above the lower epidermis is compactly arranged resembling the palisade tissue.
5. The air cavities here and there show druses.
6. Internal to the epidermis and surrounding the midrib is collenchyma.
7. Next to collenchyma is parenchyma consisting of closely packed cells.
8. The vein is surrounded by a bundle sheath and the vasculature shows the x/p arrangement.

**Xerophytic features:**

1. Presence of a thick cuticle. 2. Multiple epidermis. 3. Stomata restricted to depression of the lower epidermis. 4. Presence of trichomes in the depressions to avoid direct entry of light and wind. 5. Presence of multilayered palisade tissue.

**LEAF OF PADDY** (*Oryza sativa*):

As an example of typical isobilateral leaf, the leaf of paddy may be studied. A cross section shows the following:

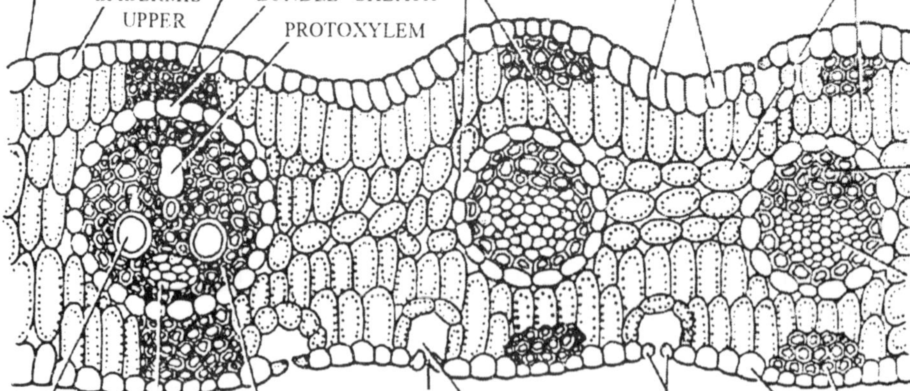


**Fig. 6.5** Leaves. T.S. of Paddy leaf

**Epidermis:**

1. There will be two epidermal layers - upper and lower.
2. The epidermal layers are uniseriate and composed of more or less oval cells arranged compactly.
3. There is the deposition of a cuticle both one the upper epidermis as well as on the lower epidermis.
4. Besides the normal cells, the upper epidermis consists of special cells called *bulliform cells* or *motor cells*.
5. These cells are larger than the other epidermal cells and are thin walled.
6. These help in the rolling of the leaves during dry weather.
7. Leaves are *amphistomatic* (stomata are present on both the epidermal layers).
8. Each stoma internally leads into a large substomatal air chamber.

**Mesophyll:**

1. It is undifferentiated.
2. But the mesophyll cells below the upper epidermis may show a compact arrangement.

**Vascular bundles:**

1. The veins are arranged parallely and are not interconnected.
2. There are two pads of sclerenchyma supporting the vein.
3. The vein itself is surrounded by a bundle sheath made up of parenchyma cells.
4. Within the bundle sheath are present collateral vascular elements showing x/p arrangement.

**LEAF OF MAIZE**

A cross section shows the following.

**EPIDERMIS:**

1. There will be two epidermal layers upper and lower.
2. The epidermal layers are uniseriate and composed of more or less oval cells arranged compactly.
3. There is the deposition of a cuticle both on the upper epidermis as well on the lower epidermis .
4. Besides the normal cells, the upper epidermis consists of special cells called bulliform cells or motor cells.
5. These cells are larger than the other epidermal cells and are thin walled.
6. These help in the rolling of the leaves during dry weather .

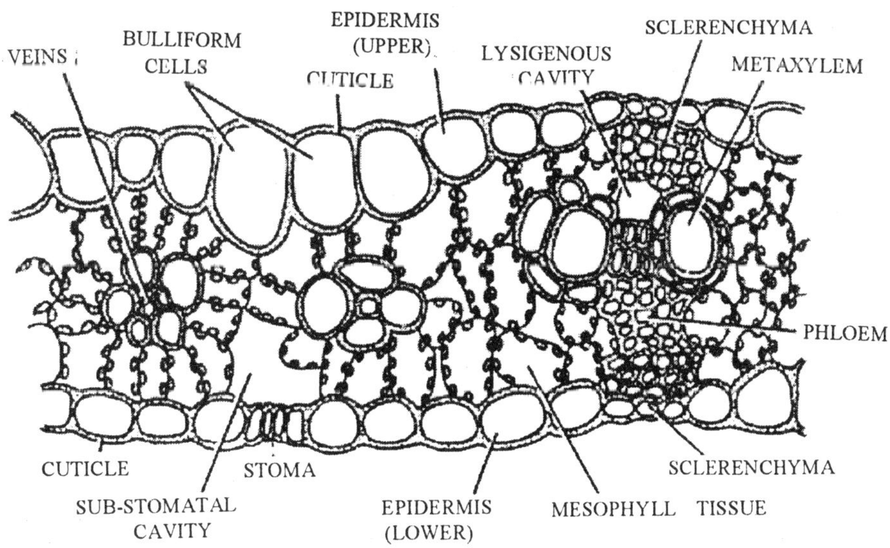


**Fig. 6.6** Leaves. T.S. of Maize leaf

7. Leaves are amphistomatic (stomata are present on both the epidermal layers ), Each stoma internally leads into a large substomatal air chamber.

**Mesophyll:**

1. It is undifferentiated.
2. But the mesophyll cells below the upper epidermis may show a compact arrangement.

**Vascular bundles:**

1. The veins are arranged parallely and are not interconnected.
2. There are two pads of sclerenchyma supporting the vein.
3. The vein itself is surrounded by the bundle sheath made up of parenchyma cells.
4. Within the bundle sheath are present collateral vascular elements showing X/P arrangement.

## STOMATA

1. Stomata are microscopic openings present in the epidermis.
2. Each stoma has a stomatal aperture or opening surrounded by two guard cells.

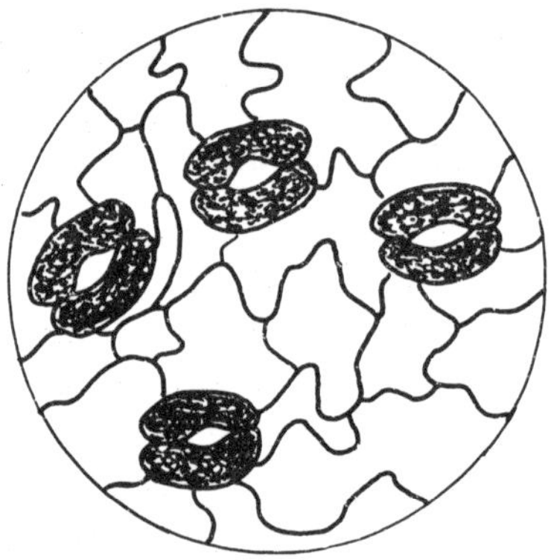

**Fig. 6.7** Anomocytic stomata (Cucurbita)

3. In the leaves, stomata may be present only on the upper epidermis (*epistomatic*), only on the lower epidermis (*hypostomatic*) or on both the epidermal layers (*amphistomatic*).
4. Guard cells are generally crescent shaped in outline and have dense granular cytoplasm with chloroplasts.
5. The cell walls of guard cells are unevenly thick and this plays an important role in the opening and closing of stomata.

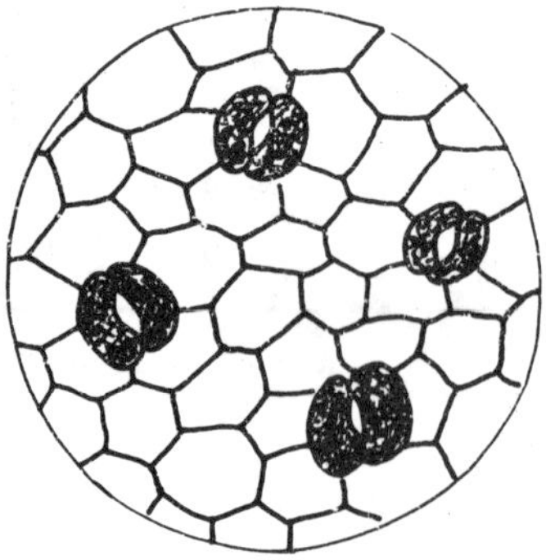

**Fig. 6.8** Anomocytic stomata (Diospyros)

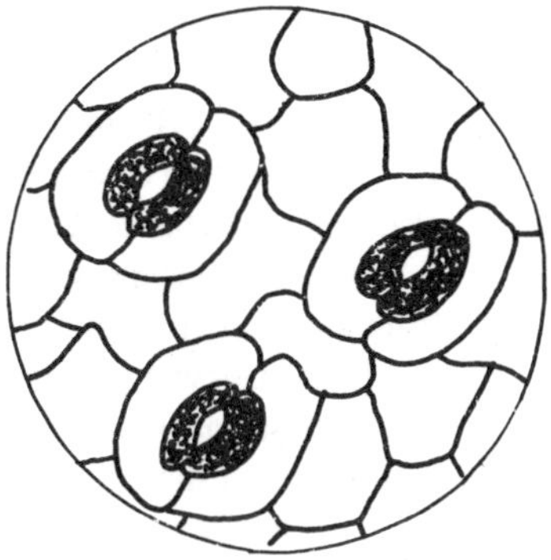

**Fig. 6.9** Paracytic stomata (Ixora)

6. Usually the wall of guard cells facing the pore is thicker than the one away from the pore.
7. During day time there is increased turgor pressure in the guard cells, as a result the thin walls are pulled away so that the concavity of the front (thick) wall increases, and the aperture becomes wider.

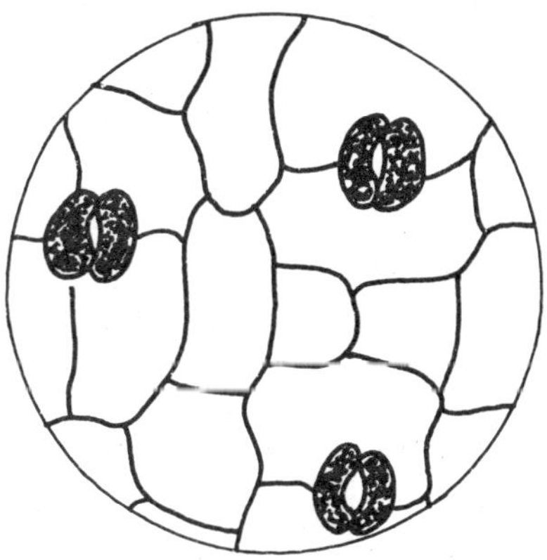

**Fig. 6.10** Diacytic stomata (Dianthus)

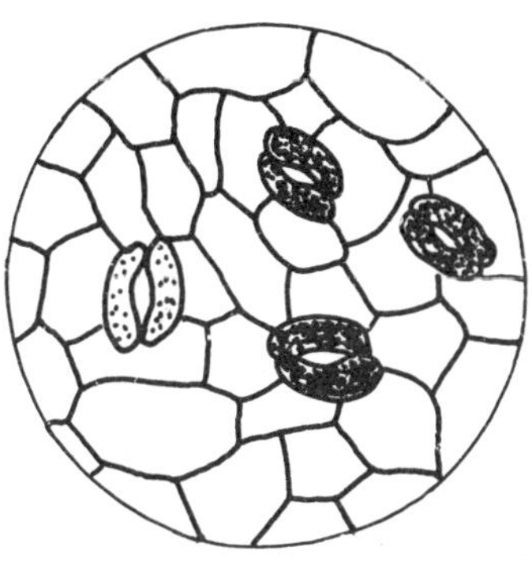

**Fig. 6.11** Anisocytic stomata (Petunia)

8. Stomata are present on the aerial organs of plants like leaf, stem, sepals etc.
9. The chief distribution of stomata however is on the leaves.
10. Rarely stomata are enocuntered on petals, stamens and ovules also, but these are non functional.

***TYPES OF STOMATA:***

1. Classification of the types of stomata is mainly based on the nature and number of *subsidiary cells* associated with them (stomata).
2. Subsidiary cells are epidermal cells that are morphologically

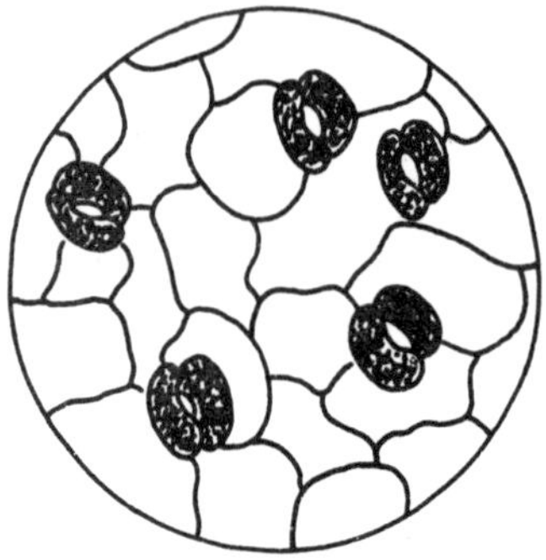

Fig. 6.12 Anisocytic stomata (Brassica)

distinct from other epidermal cells are are in close association with the stomata.

3. In Angiosperms the following main types of stomata have been identified.

**(i) Anomocytic or irregular celled type:**

1. This was first discovered in the family *Ranunculaceae*, hence also known as the *Ranuculaceous type*:
2. Here the subsidiary cells are absent in that they are not distinct from the other epidermal cells.

**(ii) Anisocytic or unequal celled type:**

1. Each stoma is surrounded by three subsidiary cells, of which one is smaller than the other two.
2. This has been reported in *Cruciferae*.

**(iii) Paracytic or parallel celled type:**

1. Stoma is surrounded by two equal parallely arranged subsidiary cells.
2. This is of frequent occurrence in *Rubiaceae*, hence also known as the *Rubiaceous type*.

**(iv) Diacytic or cross celled type:**

1. The stoma is surrounded by two subsidiary cells, which are at right angles to the guard cells unlike in the paracytic type.
2. This was first discovered in *Caryophyllaceae*, hence also known as the *Caryophyllaceous type*.

**(v) Graminaceous type:**

1. This is characteristic of *Graminae* (Poaceae) and *Cyperaceae*.
2. Here the subsidiary cells are dumb-bell shaped i.e., the middle portion is narrower than the ends and has a thicker wall.

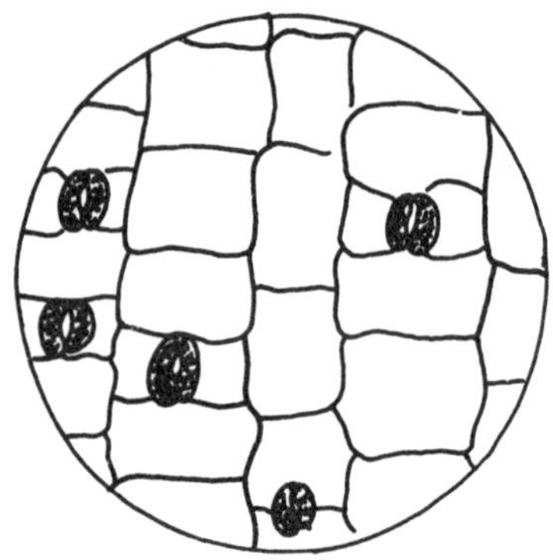

Fig. 6.13 Stomata of Rheo

Fig. 6.14 Graminaceous stomata

CHAPTER 7

# ANOMALOUS STRUCTURE OF STEMS

### STEM OF *ARISTOLOCHIA* (Aristolochiaceae)

1. The young stem has a typical dicotyledonous structure.
2. Pericycle is continous and is represented by sclerenchyma.
3. The vasculature consists of collateral bundles separated by wide interfascicular area.
4. During secondary growth inter and intrafascicular cambia become active and form a ring.
5. But while the intrafascicular cambium forms the vascular tissues, the segments of interfascicular cambium form only ray like parenchyma with the result the vascular bundles though increased in size, remain discrete.
6. With the increase in diameter the pith and rays get crushed.

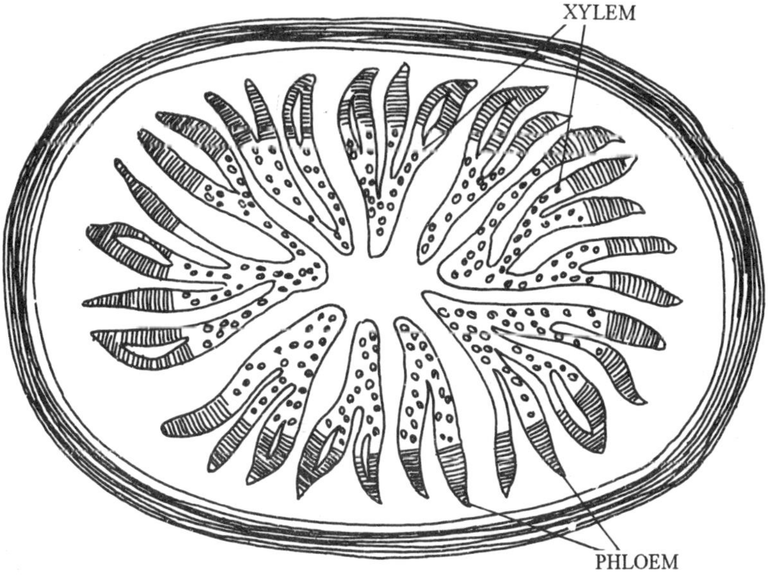


Fig. 7.1 **Anomalous secondary growth.,**
T.S. of Aristolochia stem (Diagrammatic).

7. Sometimes, portions of intrafascicular cambia also form parenchymatous rays, with the result the vascular bundles become bifurcated.
8. A stongly fluted vascular cylinder is thus formed.
9. This is an instance of the abnormal behaviour of the normal cambium.

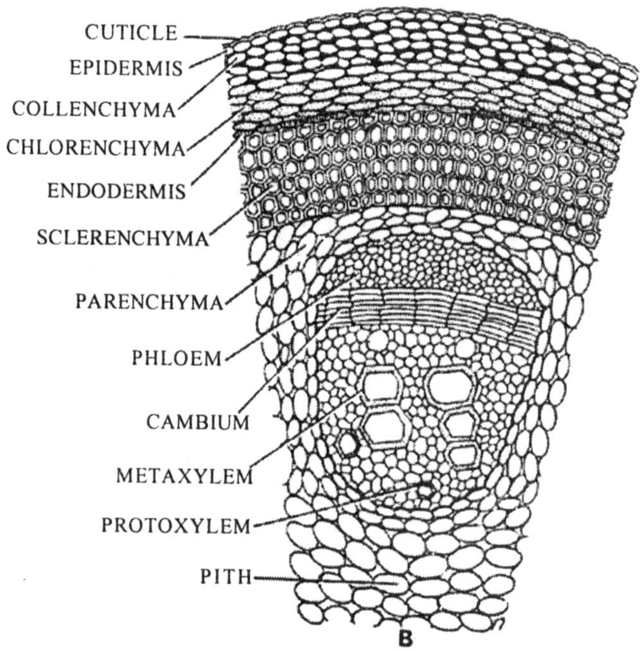


**Fig. 7.2** Anatomy of stem. T.S. of young stem of Aristolochia (dicot). detail of a sector

### *TECOMA*

1. A transverse section shows epidermis, cortex and stele in primary structure.
2. In the secondary structure the epidermis is replaced by the periderm.
3. The periderm is similar to what is seen in the stem of *Tilia*.
4. The primary cortex is mainly parenchymatous.
5. In the primary structure the cortex will have two zones - outer chlorenchyma and inner parenchyma.
6. The pericycle consists of patches of sclerenchyma.
7. In the primary structure the vascular bundles are typical to a dicotyledonus stem.
8. In the secondary structure the inter and intrafascicular cambia form a completely closed ring and cut off secondary vascular tissues.
9. As secondary growth proceeds the primary phloem gets crushed.
10. The primary xylem is found internal to secondary xylem.
11. Anomalus secondary growth is formed due to the formation of the secondary cambium internal to the primary xylem in a few places.
12. This abnormal cambium cuts off secondary xylem to the exterior and the secondary phloem to the interior.
13. In the central region is found a large pith made up of parenchyma.

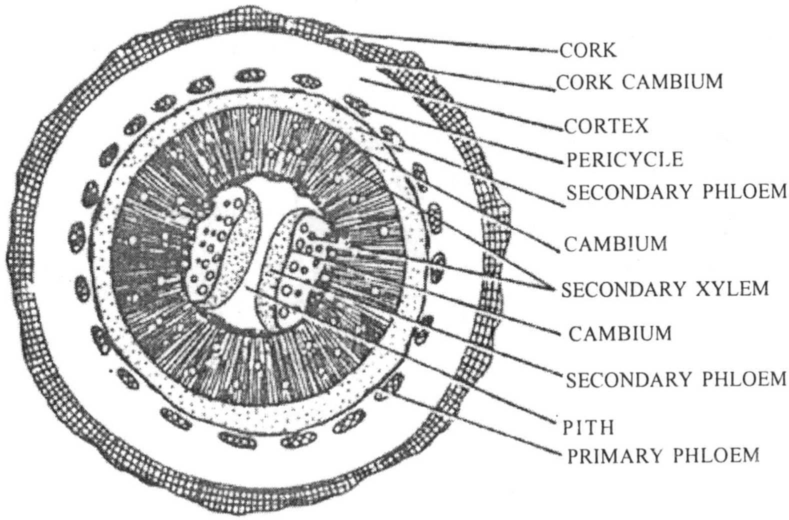


**Fig. 7.3** T.S. of Tecoma stem (diagrammetic)

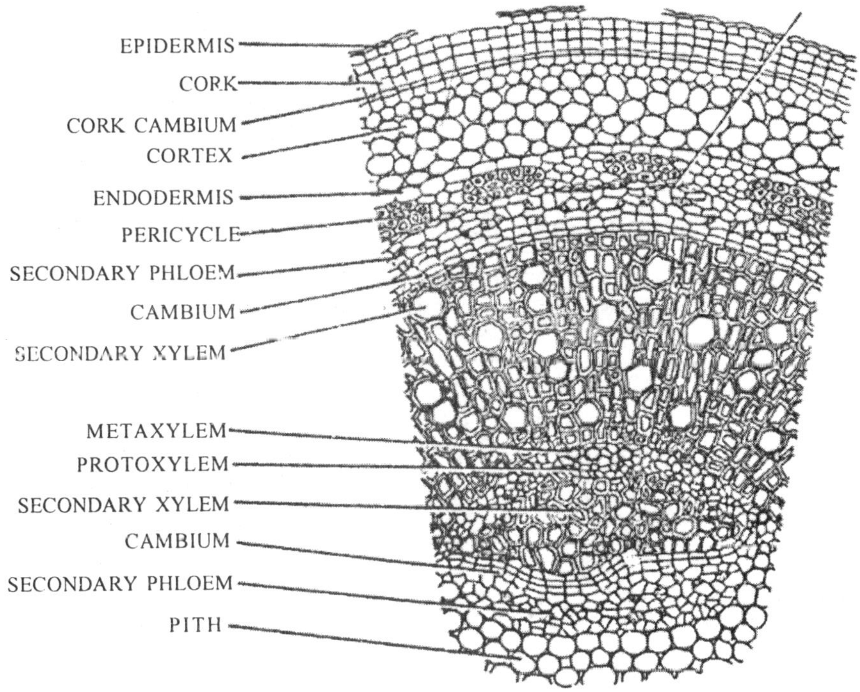


**Fig. 7.4** T.S. of Tecoma stem (a part cellular)

## STEM OF *AMARANTHUS* (Amaranthaceae)

The stem of *Amaranthus* shows certain anomalous structures, both in the primary structure as well as in the secondary thickening.

A cross section of the stem shows the following features.

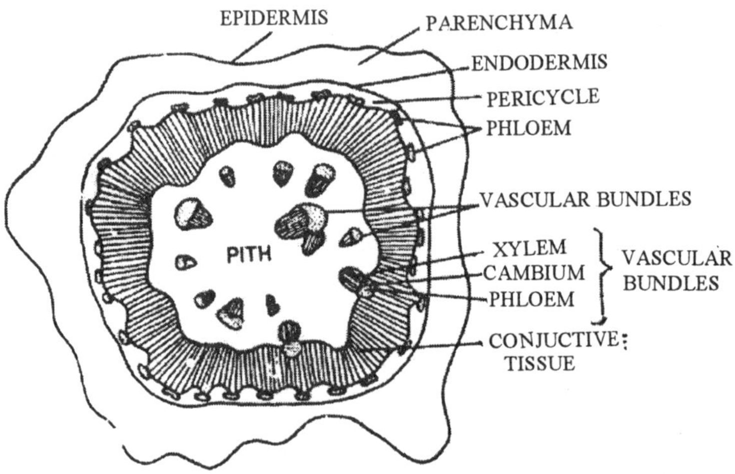


**Fig. 7.5 Anomalous secondary growth.,**
T.S. of Amaranthus stem (ground Plan).

**Epidermis:**

1. This is the outermost layer and is uniseriate.

**Ground tissue:**

1. Below the epidermis is the ground tissue made up of collenchyma, chlorenchyma and parenchyma.
2. Collenchyma is present immediately below the epidermis.

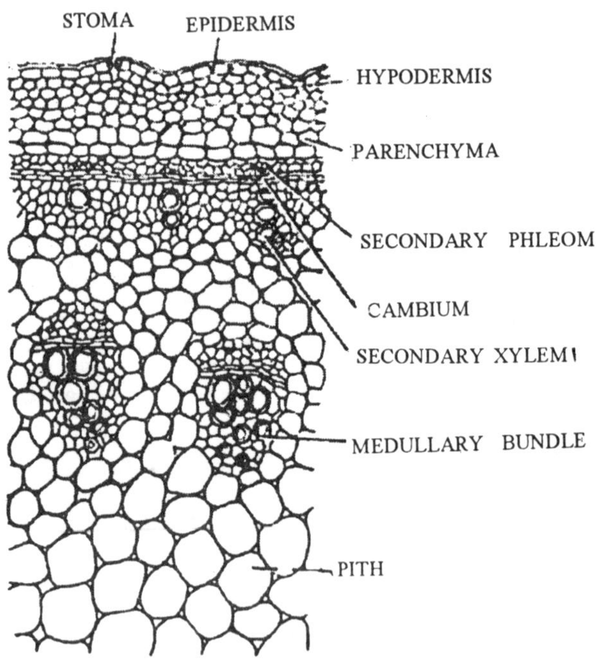


**Fig. 7.6 Anomalous secondary growth.,**
T.S. of Amaranthus stem (a portion enlarged).

3. Here and there it is interrupted by patches of chlorenchyma.
4. Below the collenchyma is a multilayered parenchymatous zone constituting the inner cortex.
5. The endodermis is not very conspicous.
6. In the primary structure the entire region inner to the endodermis is filled by loosely arranged, thin walled parenchyma cells.
7. Embedded in this, are found the scattered primary vascular bundles.
8. These are the medullary bundles.
9. There is no ring of vascular bundles in the primary structure as in a normal dicot stem.
10. The medullary bundles are conjoint, collateral, and open with endarch xylem.
11. Vascular bundles towards the periphery are smaller than those towards the centre.
12. The cambia of these bundles produce a little secondary vascular tissue (which is limited to the bundles) and very soon cease to function.
13. Secondary growth is brought about by a cambium originating outside the stele.
14. This cuts off secondary vascular bundles, which lie embedded in the conjunctive tissue around the pith region.
15. This is an example of *medullary bundles,* and *replacement of normal cambium by abnormal accessory cambium.*
16. There is anomaly both in the primary structure as well as in the secondary growth.

### STEM OF *BOUGAINVILLEA* (Nyctaginaceae)

The stem in its primary structure shows an epidermis, cortex and a vascular system. The epidermis is single layered and has many multicellular hairs.

1. **Cortex** is made up of collenchyma and chlorenchyma.
2. Collenchyma is present in the form of patches.
3. Chlorenchyma is internal to collenchyma and extends between the patches of the latter.

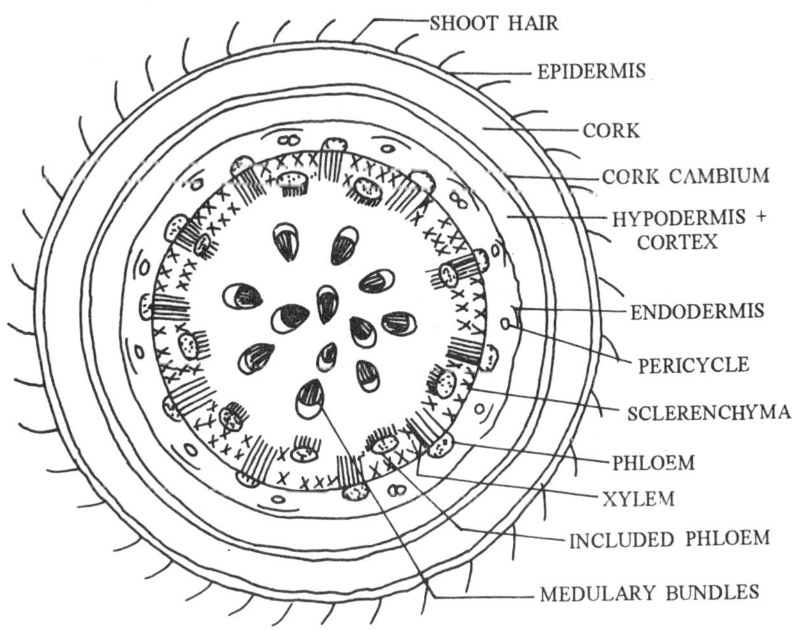


Fig. 7.7 **Anomalous secondary growth.,**
**T.S. of Bougainvillea stem (ground plan).**

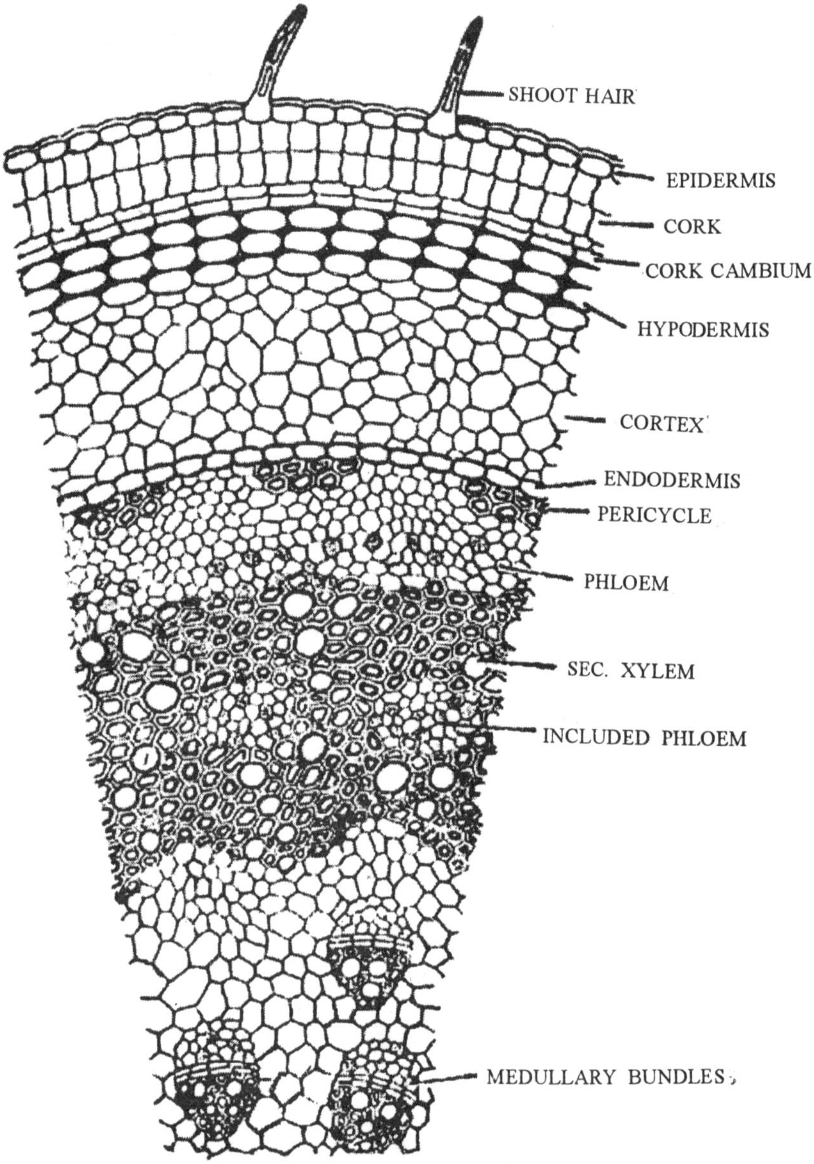


Fig.7 .8 **Anomalous secondary growth.,**
T.S. of Bougainvillea stem (A portion enlarged).

4. Internal to the cortex is an endodermis and a pericycle made up of one or two layers of parenchyma cells.
5. The vascular system consists of numerous medullary bundles embedded in the pith, in addition to the normal rings of conjoint, collateral, open endarch vascular bundles.
6. The medullary bundles are of two sizes.
7. There are four large bundles in the centre surrounded by many, smaller bundles.
8. Medullary bundles are similar in structure to that of normal bundles.
9. Fascicular cambium in these bundles has a limited activity.
10. During secondary growth, more than one cambial ring is formed.

11. The first cambial ring produces vasculature, only where the primary vasculature is present.
12. The interfascicular cambium produces living, thick walled conjunctive tissue.
13. After some time, this cambium ceases to function and a second cambial ring arises external to the first one.
14. The second cambial ring functions in the same way as the first. producing vasculature only in strips.
15. Successive cambial rings arise and function in the same way producing discrete secondary vascular bundles embedded in a sclerenchymatous conjunctive tissue.
16. The type of secondary growth here is same as in *Amaranthus* i.e., *medullary vascular bundles* and abnormal behaviour of *accessory cambia*.

### *PEPEROMIA*

1. A transverse of the section shows the epidermis, cortex and the vasculature.
2. Anomalous secondary growth is seen in the stem.
3. Epidermis is made up of a single layer of compactly arranged cells.
4. There is the deposition of cuticle above the epidermis.
5. Next to the epidermis is found the cortex consisting of few layers of collenchyma cells.
6. Next to the cortex is found a uniform mass of parenchyma constituting the ground tissue.
7. Embedded in the ground tissue are found a number of scattered vascular bundles.
8. This arrangement is unusual for dicotyledons.

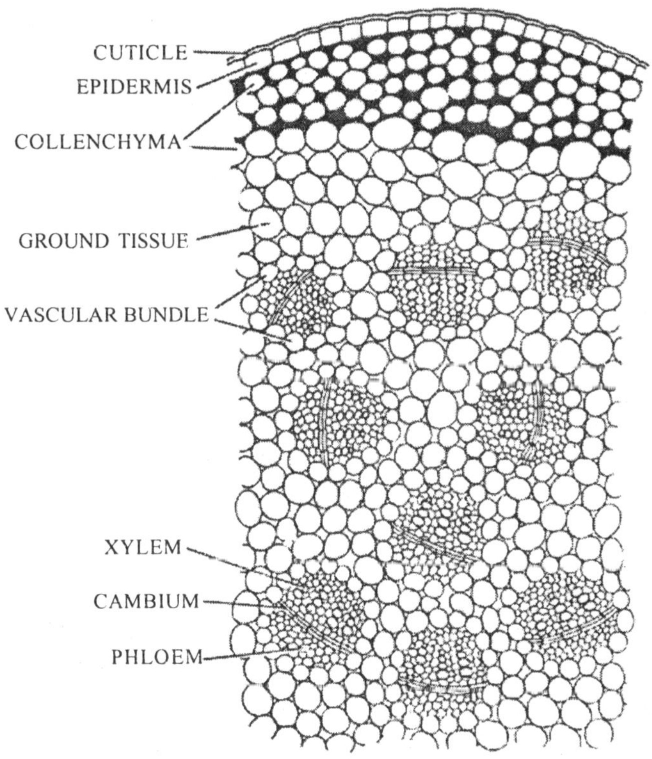


**Fig. 7.9** Anatomy of anomalour stem T.S. of Peperomia (dicot stem; diagrammetic)

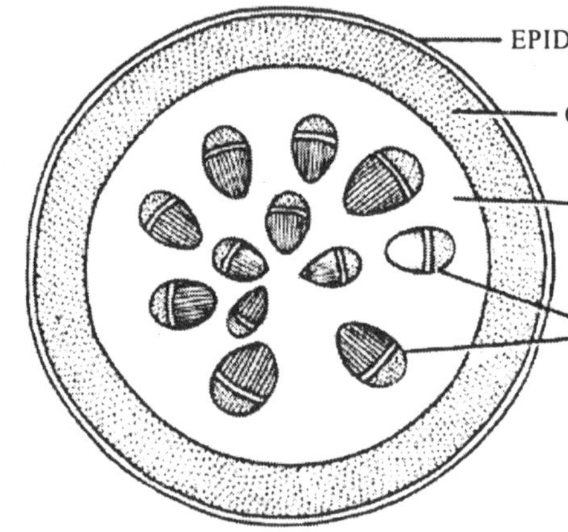


**Fig. 7.10** Anatomy of anomalous stem T.S. of Peperomia (dicot stem; detail of a sector)

## *LEPTADEMIA*

1. A transverse section of the stem show epidermis, cortex and stele.
2. Anomalous vascular structure are seen in the stem.
3. Epidermis is made up of a single layer of compactly arranged cells.
4. There is the deposition of cuticle above the epidermis.
5. Below the epidermis is found the cortex made up of chloroenchyma.
6. Sometimes below the chlorenchyma is found one or two layers of parenchyma.
7. Occasionally few stone cells may be found between parenchyma and chlorenchyma.
8. The endodermis is clearly distinguishable.
9. Next to the endodermis is found the pericycle.
10. The vasculature consists of xylem and phloem.
11. Secondary phloem is very conspicuous while the primary phloem is seen at regions opposite to the primary phloem.

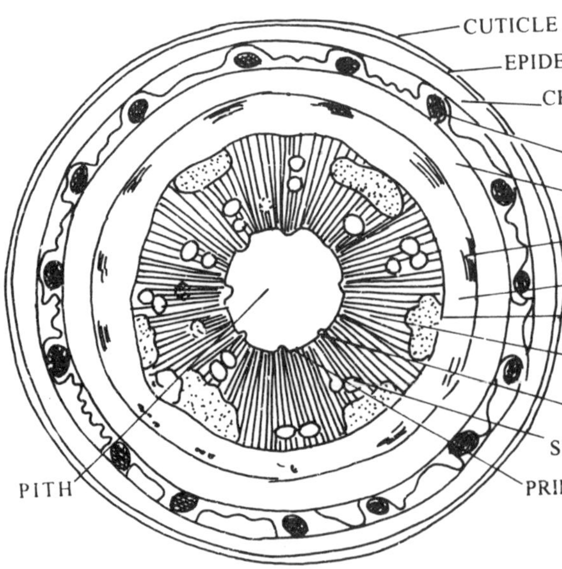


**Fig. 7.11** Anatomy of anomalous stem T.S. of Leptodenia (Aselepiadaceae -dicot) stem, diagrammatic

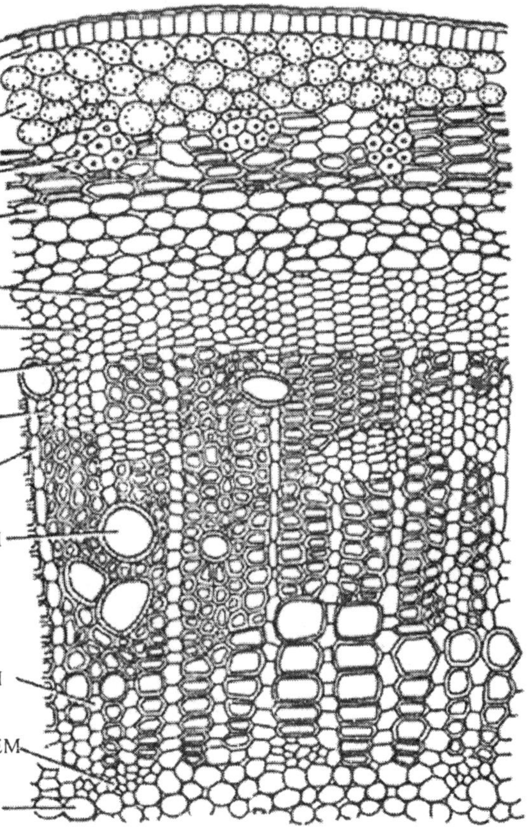


**Fig. 7.12** Anatomy of anomalous stem T.S. of Leptodenia (Aselepiadaceae -dicot) stem, detail of a sector

12. Next to the secondary phloem is the cambium made up of many layers of rectuangular cells.
13. The xylem is found next to the phloem.
14. The primary xylem is found to the interior.
15. The secondary xylem consists of large vessels.
16. The secondary vasculature has unusual feature such as interxylary and intraxylary phloem.
17. The inter xylary phloem is found between the strands of the secondary xylem.
18. The intra xylary phloem is found at the border of the pith.
19. Uni or multi seritae medullary rays are found between the vascular bundles.
20. Latex in the cells are found embedded in the parenchymatous cells.

### *SALVADORA*

1. A transverse section of the stem shows epidermis, cortex and stele.
2. Anomalous vascular structure are seen in the stem.
3. Epidermis is made up of a single layer of compactly arranged cells.
4. There is the deposition of cuticle above the epidermis.
5. Next to the epidermis is found a few layers of parenchyma.
6. Next to the parenchyma is found three to four layers of chlorenchyma.
7. In the secondary structure the epidermis is replaced by the cork tissue.

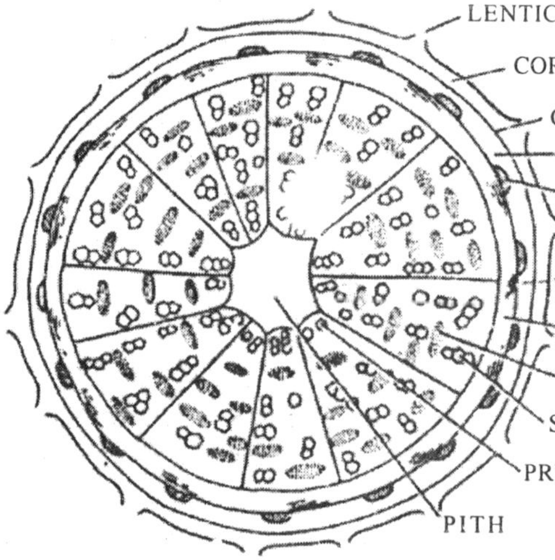


**Fig. 7.13** T.S. of Salvadora stem, diagrammatic

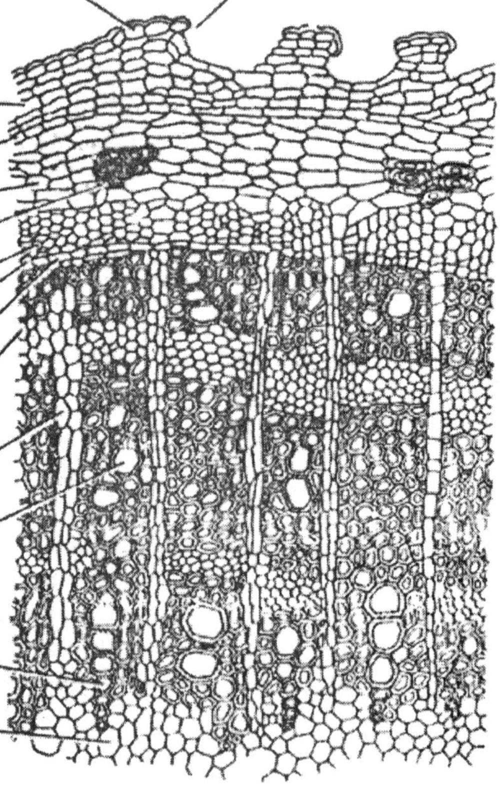


**Fig. 7.14** Anatomy of anomalous stem T.S. of Salvadora stem, Detail of a sector

8. Internal to the cork is found the cork cambium and the secondary cortex.
9. Outlining the stele is found the endodermis.
10. Pericycle is found next to the endodermis and is represented by the patches of fibers oppossite the vascular bundles.
11. Internal to the pericycle is found the secondary phloem and sum amount of primary phloem.
12. The phloem bundles is seperated by uni or multiseriate medullary rays.
13. Next to the phloem is found the cambial zone.
14. The cambium is seperated into patches by the medullary rays.
15. Next to the cambium is found the secondary xylem.
16. Towards the pith is found primary xylem.
17. The secondary xylem consists of patches of phloem constituting the interxylary phloem.
18. Towards the center is found the pith made up of parenchyma.
19. The main anomalous feature in the stem is the presence of the patches of interxylary phlcem arising on the inner side of the cambium.

### *STRYCHNOS*

1. A transverse section of the stem shows epidermis, cortex and stele.
2. Anomalous vascular structure are seen in the stem.
3. Epidermis is made up of a single layer of compactly arranged cells.
4. There is the deposition of cuticle above the epidermis.

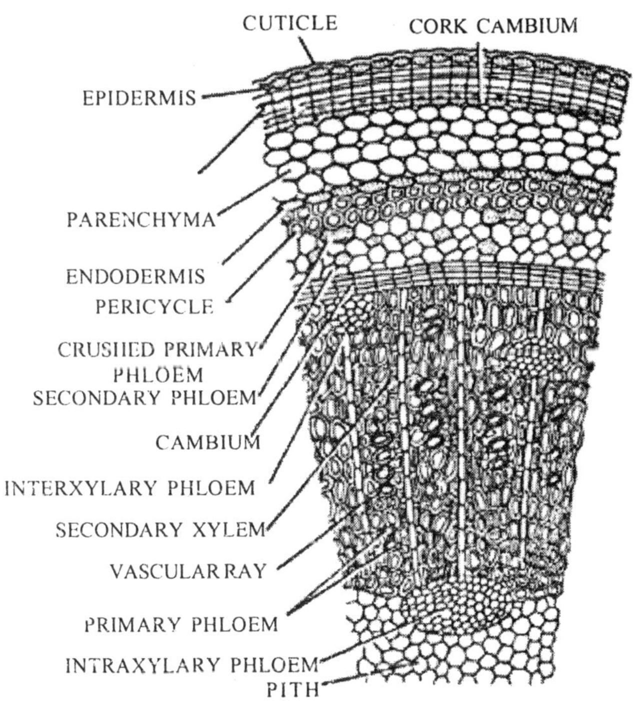


**Fig. 7.15** Anatomy of anomalous stem of Strychnos (dicot), T.S. of stem, detal of sector

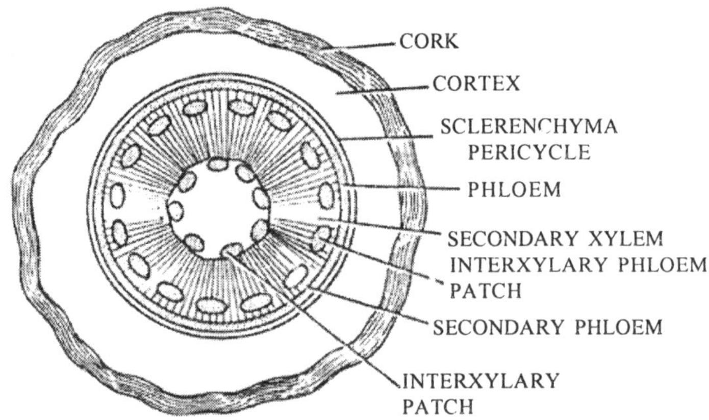


**Fig. 7.16** T.S. of Strychnos stem (dicot), diagrammatic

5. Cortex consists of a few layers of chloencyma followed by parenchyma.
6. In the secondary structure the epidermis is replaced by the cork.
7. Internal to cortex is the endodermis.
8. Next to the endodermis is found a continous layer of sclerenchyma representing the pericycle.
9. The vasculature consists of conjoint, collateral and open vascular bundles in the primary structure.
10. Secondary phloem is found near the pericycle along with cushed primary phloem.
11. Uniseriate medullary rays are present connecting the pith and cortex.
12. Secondary xylem is found next to the cambium.
13. The unusual features are presence of interxylary and intraxylary phloem.
14. Intraxylary phloem is found in the form of patches towards the pith.
15. The central region consits of a wide parenchymatous pith.

***MIRABILIS***

1. A transverse section of the stem shows epidermis, cortex and stele.
2. Anomalous vascular structure are seen in the stem.
3. Epidermis is made up of a single layer of compactly arranged cells.

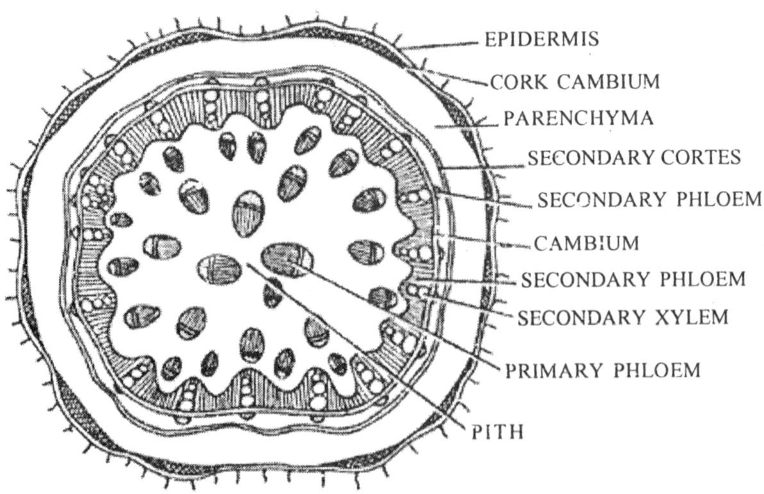


**Fig. 21.17** T.S. of stem Mirabilis, Diagrammatic

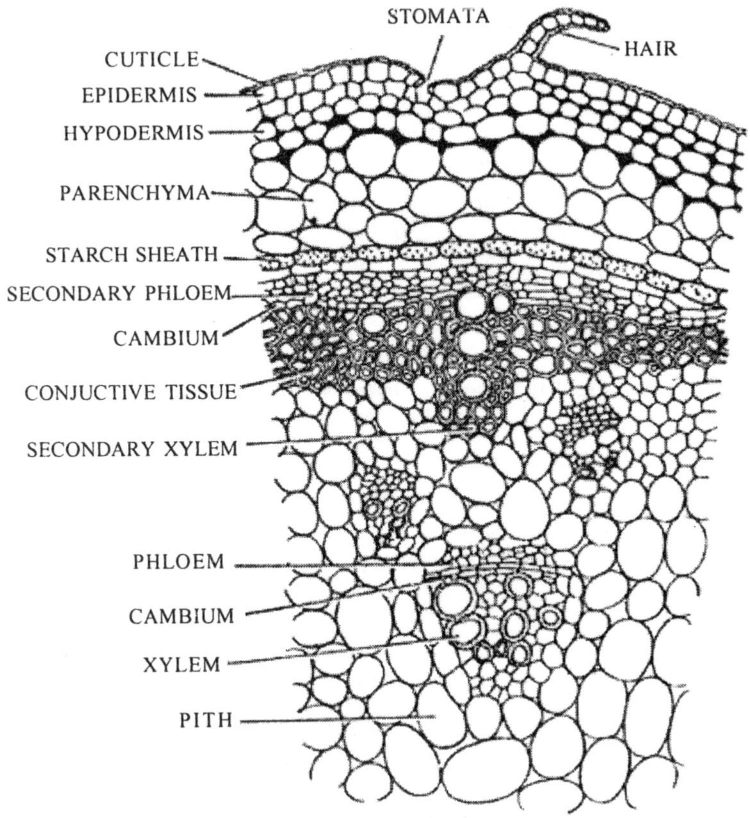


**Fig. 7.18** Anatomy of anomalous stem of Mirabilis, Detail of sector of same

4. There is the deposition of cuticle above the epidermis.
5. Cortex consists of few layers of chloenchyma followed by parenchyma.
6. Internal to the cortex is the endodermis.
7. Internal to the endodermis is found pericycle made up of a single layer of parenchyma.
8. Next to the pericyle is the secondary phloem along with patches of primary phloem.
9. Next to the phloem is the cambium.
10. Internal to the cambium is the secondary xylem.
11. The central region of the stem consists of a uniform mass of parenchyma constituting the ground tissue.
12. Embedded in this ground tissue are embedded a number of vascular bundles which are reffered to as the medullary vascular cells.
13. Anomalous structures are seen in the form of a number of rings of vascular bundles occuring in the medullary region.
14. These vascular bundles are irregular arrangement.

### *ARGYREIA*

1. A cross section of the stem is some what rectangular with four projections at the corner.
2. In the primary structure the usual epidermis, cortex and stele are seen.
3. Epidermis is single layered made up of compactly arranged cells.

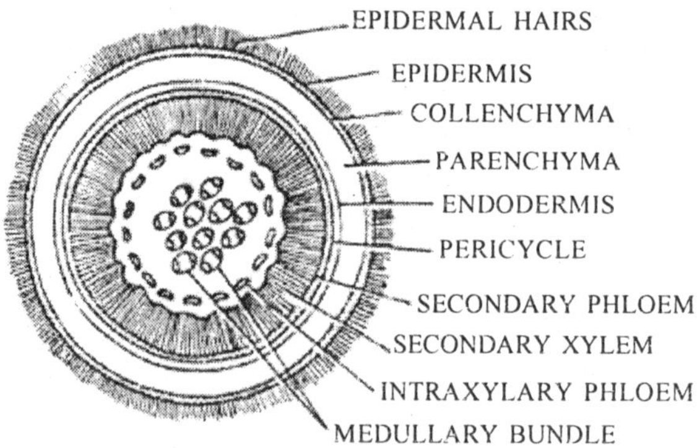


**Fig. 7.19** Anatomy of anomalous stem of Argyreia Diagrammatic

4. Cuticle is deposited above the epidermis.
5. The cortex consists of outer collenchyma and inner parenchyma.
6. The cortex is followed by endodermis.
7. Next to the endodermis is the single layered parenchymatous pericycle.

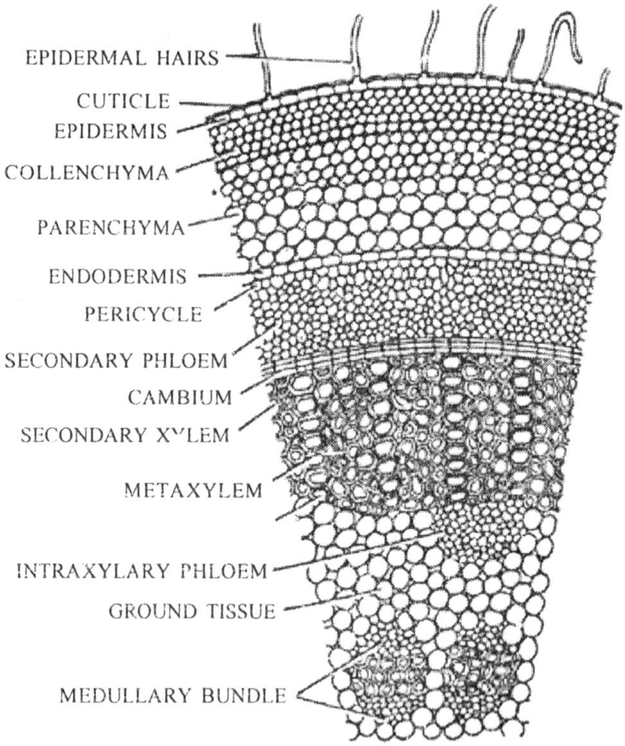


**Fig. 7.20** Anatomy of anomalous stem of Argyreia, Detail of a sector of same

8. The vasculature consists of conjoint, colletral and open vascular bundles in the primary structure.
9. Next to the pericycle is the secondary vasculature consisting of secondary phloem, cambium followed by secondary xylem.
10. Central region is occupied by the pith.
11. The unusal feature is the presence of several discrete vascular bundles found embedded in the pith.
12. Another unusal feature is the presence of patches of interxylary pholem.

### *CHENOPODIUM*

1. A cross section of the stem is some what rectangular with four projections at the corner.
2. In the primary structure the usual epidermis, cortex and stele are seen.
3. Epidermis is single layered made up of compactly arranged cells.
4. Cuticle is deposited above the epidermis.
5. The cortex is divided into two zones - outer collenchymatous and inner parenchymatous.
6. Internal to the cortex is the endodermis.
7. A parenchymatous pericylce is found internal to the endodermis.
8. The primary vascular bundles are typical to a dicotyledonous stem.
9. In the region of the pericycle develops an unusual cambium external to primary phloem.
10. It is this which bings about the abnormal secondary growth.
11. Normal seconday also occurs in the primary vascular bundles due to the activity of fasicular cambium.
12. This cambium however is active only for a short period.

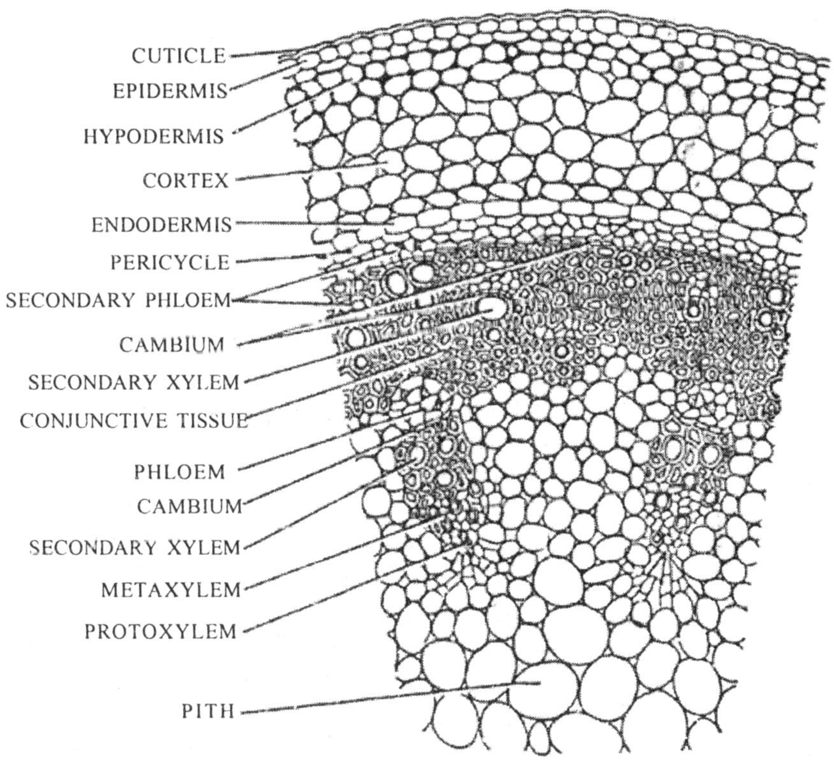


**Fig. 7.21** T.S. of Chenopodium stem (a part cellular)

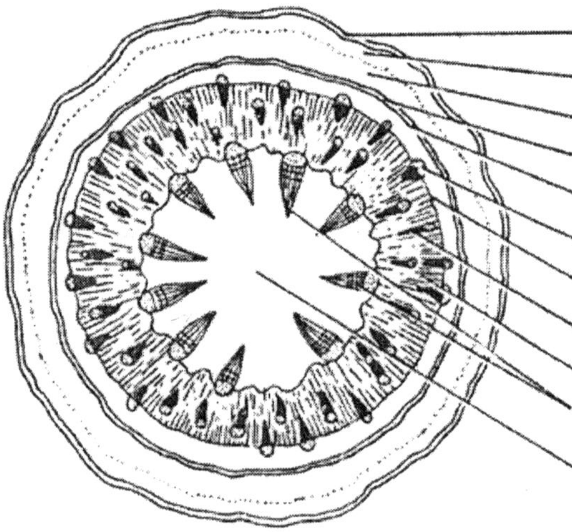


**Fig. 7.22** T.S. of Chenopodium stem (diagrammetic)

13. The abnormal cambium ring produces thin walled cells towards the inner face increasing the quantity of ground tissue.
14. As a result of this the primary vascular bundles are pushed into the centre appearing like medullary vascular bundles.
15. Towards the centre of the stem is found a large parenchymatous pith.
16. The unusual features of the stem are the presence of extrastelar cambium and medullary vascular bundles (these are normal).

***BIGNONIA***

1. A transverse section of the stem shows the following features.
2. In the primary structure the stem shows epidermis cortex and Stele.
3. The secondary growth shows both normal and anomalous structures.
4. The outline of the section will not be exactly circular.

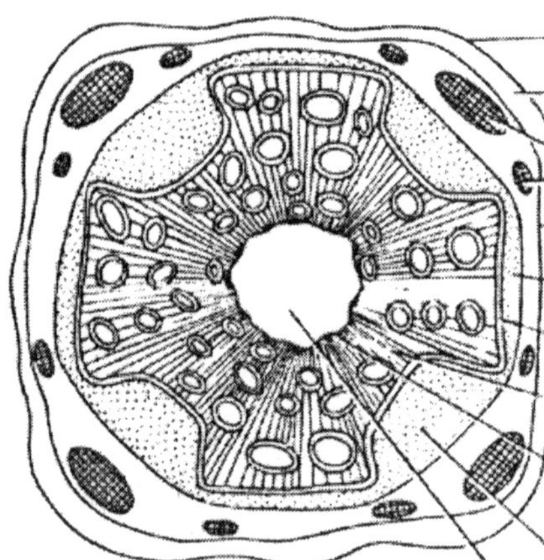


**Fig. 7.23** The stem, anomalous structure. T.S. of Bignonia (dicot). detail of the structure

5. Epidermis is made up off a single layer of cells and is covered by a thick cuticle.
6. In older stems however the epidermis is replaced by the periderm.
7. The periderm is made up of Cork, cork cambium and secondary cortex.
8. The cortex is made up of the a few layers of parenchyma.
9. The endodermis is not distinct and is made up of barrel shaped cells.
10. Pericycle is represented by patches of sclerenchyma.
11. Secondary growth takes place by the activity of normal fascicular cambium.
12. The normal cambium behaves abnormally in that it cuts off more secondary phloem to the exterior and less secondary xylem towards the interior in three or four places while in other regions it cuts off more xylem towards the interior and less phloem towards the exterior.
13. As a result the secondary vascular region becomes wavey.
14. Due to this abnormal activity in many places the secondary phloem becomes 'U' shaped and projects into secondary xylem.
15. Some of the cells in the phloem become fibrous.
16. Occupying the centre region is found the pith.

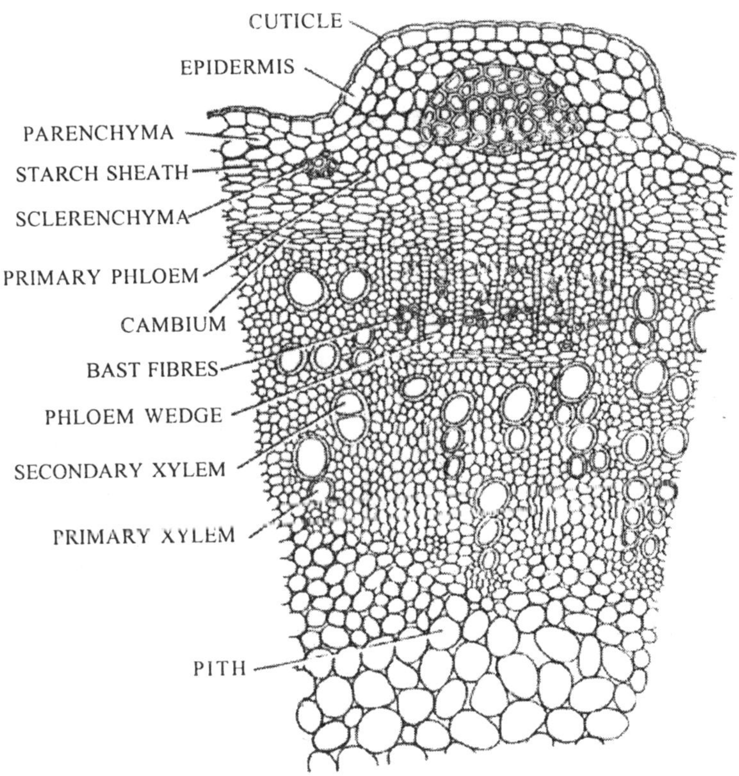


**Fig. 7.24** The stem, anomalous structure. T.S. of Bignonia (dicot). diagrammetic

## *NYCTANTHES*

1. A cross section of the stem is some what rectangular with four projections at the corner.
2. In the primary structure the usual epidermis, cortex and stele are seen.
3. Epidermis is single layered made up of compactly arranged cells.

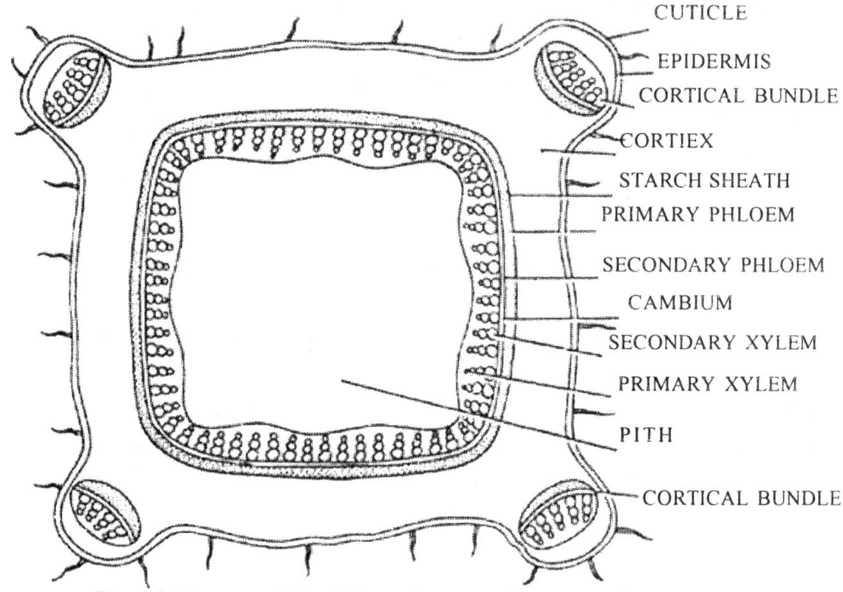


**Fig. 7.25** Anatomy T.S. of Nyctanthes stem, Diagrammatic

4. Cuticle is deposited above the epidermis.
5. Next to the cortex is found few layers of collenchyma followed by a few layers of parenchyma.
6. As the stem as four ridges the exten of the parenchyma is more at these regions.

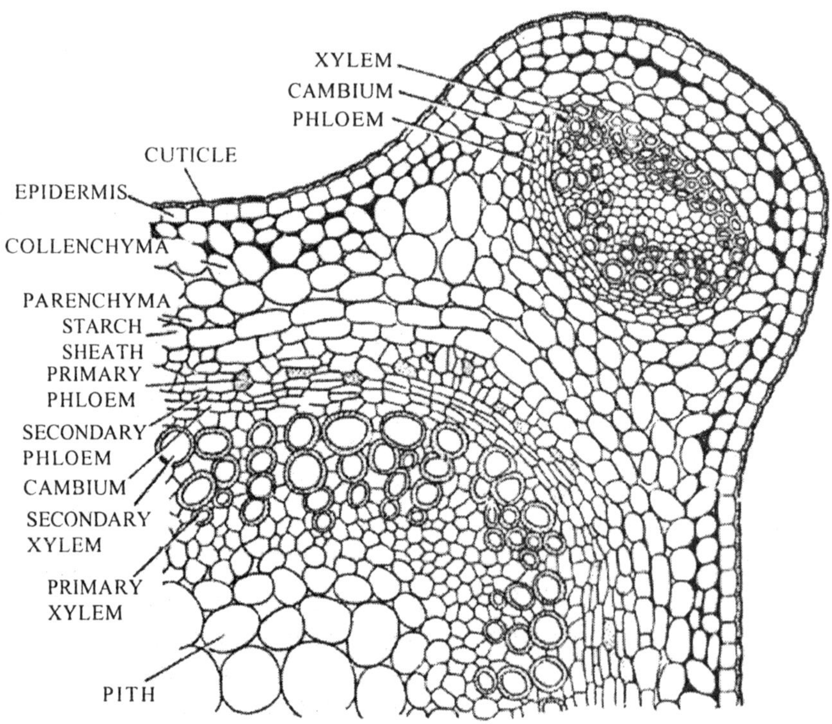


**Fig. 7.26** Anatomy. T.S. of Nyctanthes stem, sector enlarged

7. Embedded in the parenchyma at the ridges are found four cortical vascular bundles.
8. These cortical vascular bundles are anomalous in origin.
9. These cortical vascular bundles are colletral and open.
10. Next to the cortex is the endodermis.
11. Endodermis is followed by pericycle and secondary vascular tissues which are normal.
12. Cambium is found between the secondary pholem and secondary xylem.
13. Occupying the secondary region is found the pith.

### *BOERHAVIA STEM*

1. *Boerhavia diffusa* is a small semiprostrate spreading herb.
2. A transverse section of the stem shows an outermost single layered epidermis.
3. The epidermis bears a number of multicellular hairs.
4. In the primary structure, next to the epidermis internally is the hypodermis.
5. The hypodermis is made up of 2-3 layers of collenchyma which is followed by parenchymatous cortex having few chlorophyllous cells.
6. Next to the cortex is an indistinct endodermis lining externally 1-2 layered pericycle.
7. The stele consists of 3 rings of vascular bundles of which inner ring has two large centrally located medullary bundles, the middle ring is made up of 6-14 loosely arranged bundles and the outer ring has small 15-20 bundles.
8. Secondary growth is limited in the inner and middle rings.
9. As a result they slightly increase their size.
10. The inter and intrafascicular cambial strips join together to form the cambial ring.

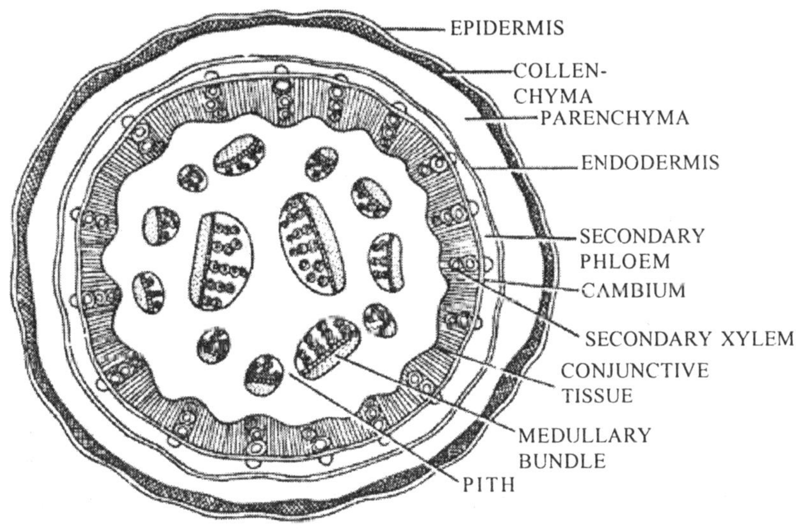


**Fig. 7.27** Boerhavia diffusa. T.S. stem-diagrammetic

11. The cambial cells produce secondary xylem in the infrafascicular region and lignified conjunctive tissue in the interfascicular region on the inner side and on the outer side secondary phloem facing the secondary xylem and parenchyma opposite the conjunctive tissue.
12. After the formation of secondary tissue, the cambium ceases its activity and a new cambial ring gets differentiated by combining the secondary parenchyma cells opposite to the conjunctive tissue and the cells of pericycle outside the phloem.
13. This accessory cambium functions in a manner similar to the fascicular cambium and produces similar tissues.

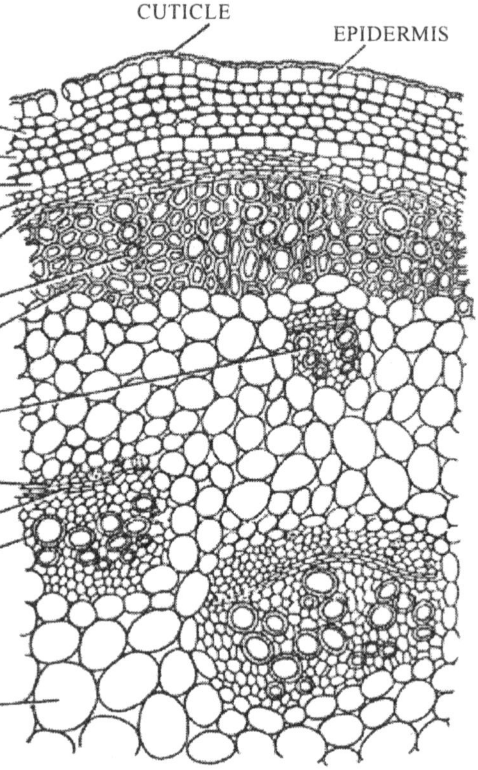


**Fig.7. 28** Boerhavia diffusa. T.S. of a sector of enlarge

14. As a result of this a fourth ring of vascular bundles of secondary origin are produced.
15. Very soon this cambium also stops functioning and one more cambium gets differentiated, which also functions in a similar manner.
16. As a result of these successive cambial differentiation several concentric rings of vascular bundles are formed.
17. These vascular bundles get embedded in the thick conjunctive tissue seperated by him walled parenchyma give the appearance of growth rings.

### *STEM OF DRACENA (Liliaceae)*

1. Generally, in monocots ther will be no secondary growth due to the lack of cambium.
2. Abnormally, secondary growth occurs in members like *Dracena, Yucca* etc.
3. The stem of *Dracena* in its primary structure consists of a normal monocotyledonous structure.
4. It has a number of conjoint, collateral and closed vascular bundles scattered in the ground tissue.
5. Secondary growth is initiated by a multilayered cambium arising outside the primary vascular bundles.
6. This abnormal cambium produces more cells towards its inner face than towards the outer face.
7. The outer derivatives develop into thin walled parenchyma whereas the inner derivatives develop partly into lignified conjunctive parenchyma and partly into secondary vascular tissues.
8. Each cambial cell divides anticlinally to form a row of four cells.

9. Further division are irregular.
10. The cells towards the periphery develop into xylem and those towards the centre develop into phloem.
11. After secondary growth, the stem of *Dracena* shows the following features.
12. Outermost layer is the epidermis.
13. Beneath the epidermis is the cork tissue.
14. Internal to this is the parenchymatous cortex which is secondary in origin.

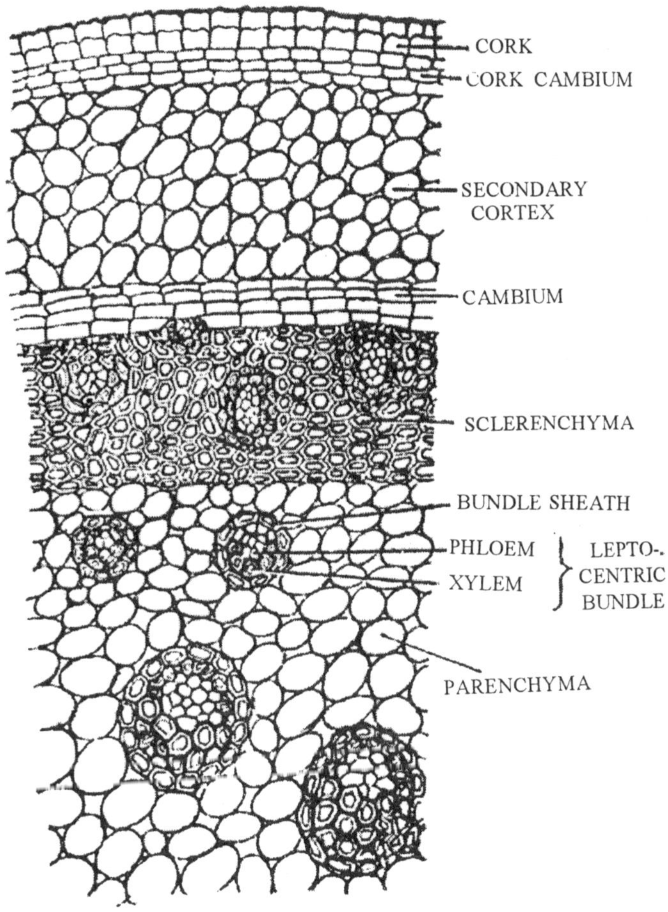


**Fig. 7.29 Anomalous secondary growth.**
T.S. of Dracena stem (A portion enlarged).

15. It is derived from the cambium.
16. Next to the secondary cortex is the cambium.
17. Internal to the cambium are found the secondary vascular bundles embedded in the thick walled conjunctive tissue.
18. The secondary vascular bundles are *amphivasal* (phloem surrounded by xylem on all the sides).
19. Towards the centre of the stem are found the normal primary vascular bundles.

## STORAGE ROOT OF BETA *VULGARIS* (Beet root)

1. The storage roots generally represent the combination of hypocotyl and root and exhibit a great variation in their secondary structure.

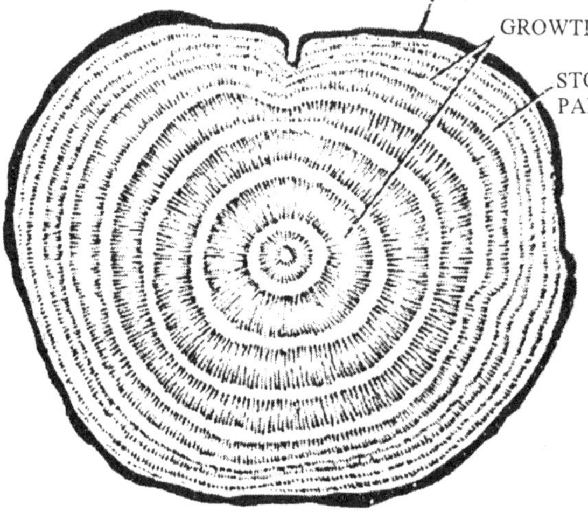


**Fig. 7.30 Anomalous secondary growth.,**
T.S. of storage root of Beta vulgaris (ground plan).

2. The storage root of *Beta vulgaris* may be taken as an example of abnormal secondary thickening in the storage roots.
3. According to Artschwager the young root is diarach.
4. At the intiation of secondary growth, the primary cambium arises between the xylem and phloem groups, except opposite to the protoxylem and phloem groups, where it arises from the pericycle.
5. This produces the innermost ring of vascular bundles.
6. Very soon, this cambium ceases to function.
7. Second cambium arises from the phloem parenchyma outside the first ring of primary cambium.
8. The secondary cambium produces another ring of vascular bundles.
9. This also ceases to function after sometime.
10. The next cambial ring is formed in the pericycle.
11. The pericycle cells proliferate and produce alternating layers of vascular bundles and parenchyma cells of pericycle.
12. In this way a number of vascular rings are formed.
13. The vascular bundles are collateral and are separated by narrow bands of radial parenchyma, rich in reserve food.

# PART TWO
# EMBRYOLOGY

CHAPTER 8

# FLOWER AND ORGANOGENY

**Structure of a typical flower:** Basically, a flower consists of three parts viz., *receptacle, sterile appendages* and *fertile appendages*.

Receptacle represents the expanded or swollen tip of the stalk(pedicel) of the flower. It is on this that the appendages of the flower are borne. The receptacle may be flat, globose or concial in shape.

The sterile appendages (accessory organs) are of two types viz., *calyx* and *corolla*. Sometimes the sterile appendages cannot be distinguished into calyx and corolla as in *Michelia*. In such cases the sterile appendages are termed *perianth lobes* or *tepals*.

Calyx is the outermost or lower most of the sterile appendages. The individual members are called sepals and they are generally green in colour. The sepals in flower may be fused or free.

Corolla is next to calyx. It is the most attractive of the floral organs. The various colours of the flowers are due to corolla. The individual members are called *Petals*. The petals in the corolla may be fused or free.

In some flowers, the sterile appendages are entirely absent. Such flowers posses only the fertile appendages and are called *achlamydeous*.

The fertile appendages are also of two types viz., *Androecium* and *Gynoecium*.

**Androecium** represents the collection of male reproductive structures. In arrangement, it is found next to corolla. The individual members are called *stamens*. Each stamen has three parts viz., *filament, anther* and *connective*. Filament is stalk like and it bears the anther. The portion of the filament that joins the anther is called the connective. It is in the anther (Microsporangium) that the pollen grains are formed which ultimately develop into the male gametophyte.

**Gynoecium** represents the collection of female reproductive structures. It is also called pistil and has three parts viz., *ovary, style* and *stigma*. Ovary is the basal swollen part of the gynoecium and is made up of *carpels* (megasporophylls). Within the carpels and attached to the plaenta are found the *ovules* (megasporangia). It is within the ovule, the embryo sac (female gametophyte) develops. Further developments i.e., fertilization, endosperm and embryo also take place in the ovule. Style is a stalk that arises from the ovary and it facilitates the growth of the pollen tube. The style terminates in a stigma which serves as a lodging for pollen grains to settle and germinate.

**Organogeny:** The development of floral parts is called organogeny. The floral primordium arises on the inflorescence axis (peduncle). First, bract arises as a lateral projection in the axil of which the flower originates. The floral shoot subsequently becomes flattened and forms the receptacle. The receptacle at this stage shows a group of meristematic cells with dense cytoplasm and conspicuous nucleus. The various floral organs originate from the receptacle either *spirally* or *cyclically*. In the spiral type, the recptacle shows a tendency towards indeterminate growth in a linear fashion and assumes a dome like shape. This type is regarded primitive and is seen in *Magnoliaceae*. In the cyclic type, the floral parts arise in distinct whorls. This is seen in the majority of flowering plants. An intermediate condition where the calyx and corolla are spiral, with cyclic androecium and gynoecium is seen in the *amentiferae* group.

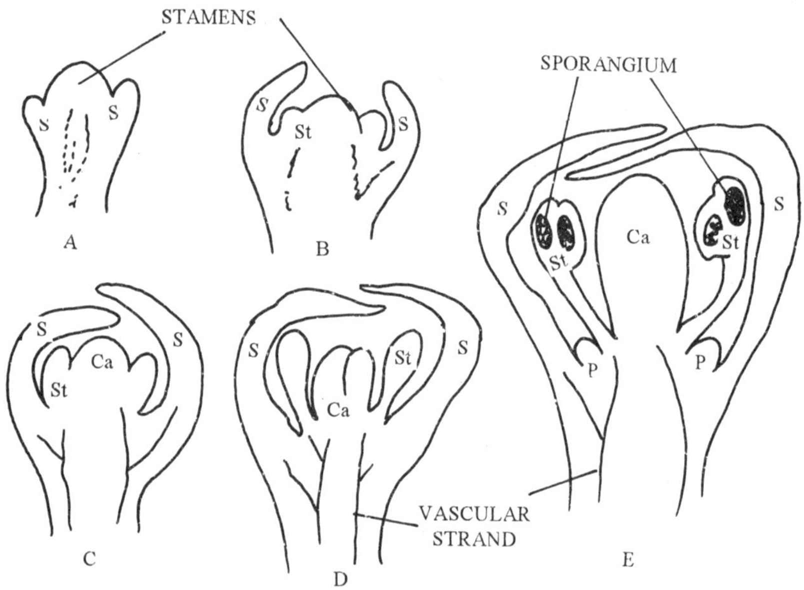


**Fig. 8.1 Flower**
**Stages in the organogenesis of a typical flower**
**(S= sepal; St= stamen; P = petal ; Ca = carpel)**

In the axil of the bract a conspicuous bulb like outgrowth arises and constitutes the floral primordium. By further growth, the floral primordium elongates and gives rise to two distinct regions viz., *pedicel* (stalk) and the *receptacle*. It is on the latter that the floral organs arise. In sessile flowers, the pedicel is suppressed

The floral parts viz., K (calyx), C(corolla), A(androecium and G(gynoecium) arise in acropetalous sucession on the floral receptable. First to arise is calyx and it is followed by C.G. and A. Variations in the organogeny are met with in several members. In *Rubus* the development is K, A, G, and C while in *Rudbeckia* it is C, A, G and K.

CHAPTER 9

# ANTHER AND MALE GAMETOPHYTE

**Development of the microsporangium:** A very young anther in a cross section shows a homogenous mass of cells bound by an epidermis. During further growth the anther assumes a four lobed shape. In each lobe a few of the hypodermal cells (sometimes one) become differentiated from the others by their prominent nucleus and dense cytoplasm. These are the *arhesporial cells*. In majority of the members the archesporium consists of several vertical rows of cells which appear as a plate of cells in a transverse section.

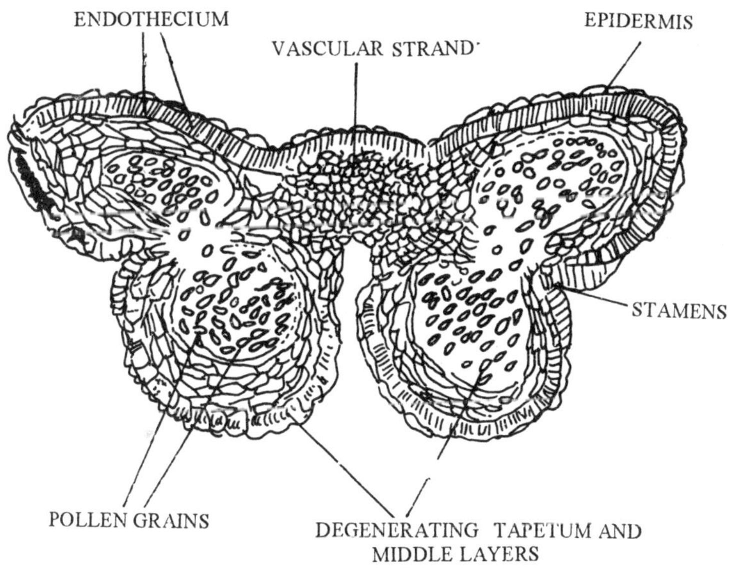


Fig. 9.1 **Microsporangium**

**Transverse section of a mature anther at the time of dehiscence**

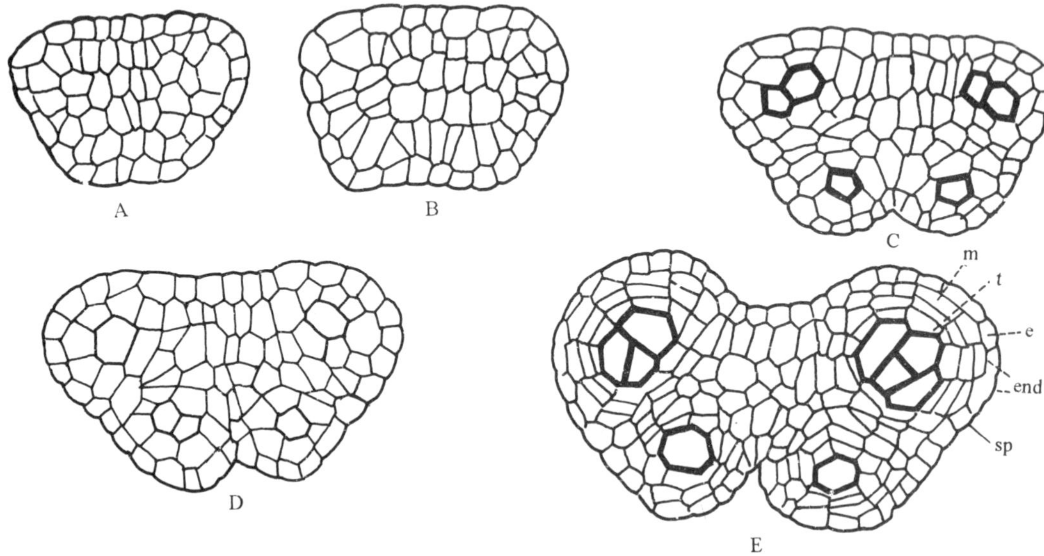


**Fig.9.2 Microsporangium**

Developmental stages of the anther (e = Show labelling on the diagram itself (redrawn from Warning, 1873)

The archesporail cells divide periclinallay cutting off *primary parietal cells* towards the epidermis and *primary sporogenous cells* towards the interior. The primary parietal layer divides to form two *secondary parietal layers*. Of these, the inner one distinguishes itself into *tapetum* while the outer divides again periclinally to form two layers of cells. Of these, the outer one forms the *endothecium* while the inner one forms the *middle layer*. The sporogenous cells develop into microspore mother cells.

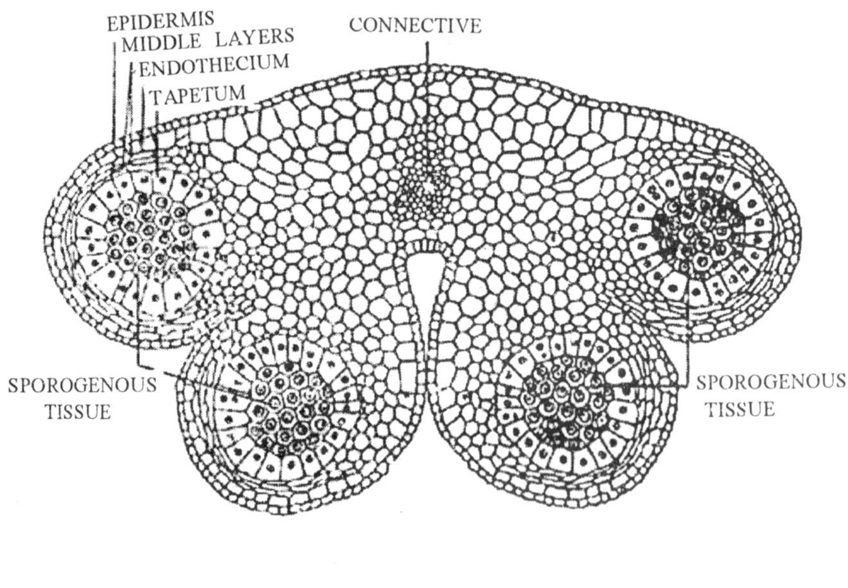


**Fig. 9.3 Microsporangium**

Transection of a fully developed tetrasporangiate anther

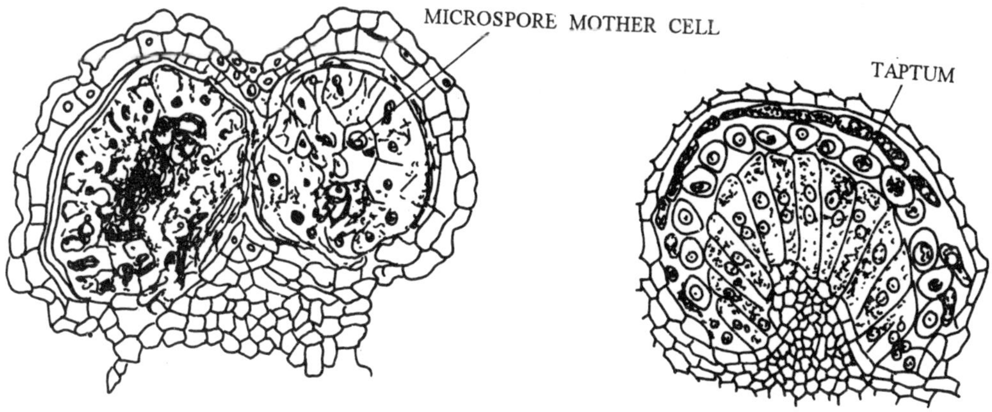


Fig. 9.4 Microsporangium. T.S. of Anther showing PMCs and tapetum

**Anther Wall:** A transverse section of a fully developed anther shows a group of microspore mother cells surrounded by a tapetum, a middle layer, an endothecium and an epidermis.

**Epidermis:** It is the outermost wall layer of the anther. During anther development, it undergoes anticlinal divisions to keep pace with the enlargement of the inner tissues. The epidermal cells become greatly stretched and flattened. In a mature anther, only the remanants of the epidermis are seen.

**Endothecium:** The layer of cells lying immediately below the epidermis constitute the endothecium. Endothecium attains its maximum development at the time of the anther dehiscence. Cells of this layer

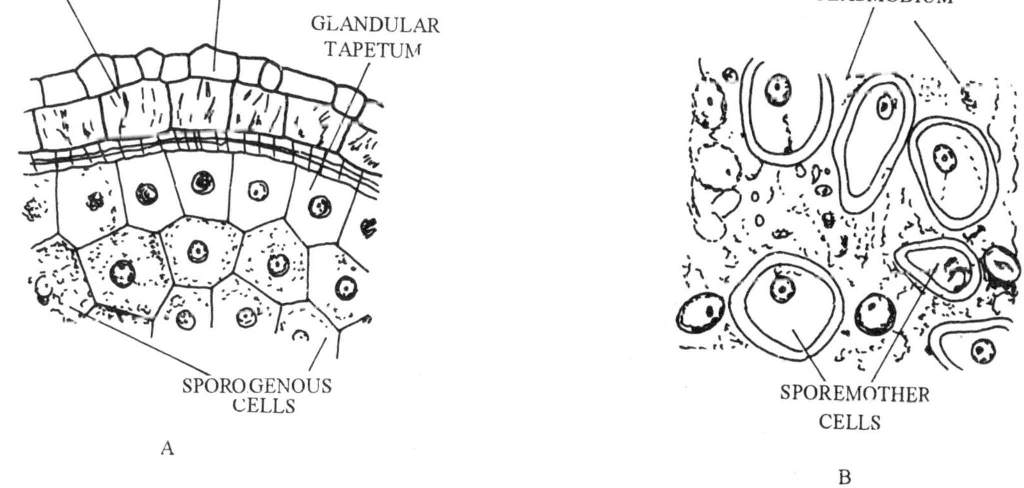


**Microsporangium**

Fig. 9.5 Types of Tapteum A. Glandular, B. Amoeboid

develop fibrous bands of callose which arise from the inner tangential walls (rarely from the radial walls) and run outward and upward ending near the outer wall of each cell. The outer tangential walls however remain thin.

These thickenings are apparently associated with the dehiscene mechanism of the anther. In certain members of *Hydrocharitaceae*, fibrous thickenings are absent. On the other hand in some members like *Argemone, Crinum* etc., one or more of the middle layers may acquire fibrous thickenings. In members of *Orobranchaceae*, some epidermal cells themselves may develop fibrous thickenings.

**Middle layers:** Usually only one middle layer is present. But, depending upon the type of anther wall development there may be one (dicotyledonous and monocotyledonous types), two (basic type) or no (reduced type), middle layers. Occassionally as in *Melastoma malabathricum*, the number of middle layers may be seven. Cells of the middle layer lack the capacity to divide anticlinally and hence the tissue cannot adjust itself to the mutiplying and expanding sporogenous cells within the sporangium. As a result, it gets crushed and cannot be seen in the mature anther. The formation of middle layers is regarded as a relic feature. In some species, on the other hand (*Ranunculus, Lillium* etc.), the middle layers persist and act as storage centres for starch and other reserves which get mobilised during later development of the wall.

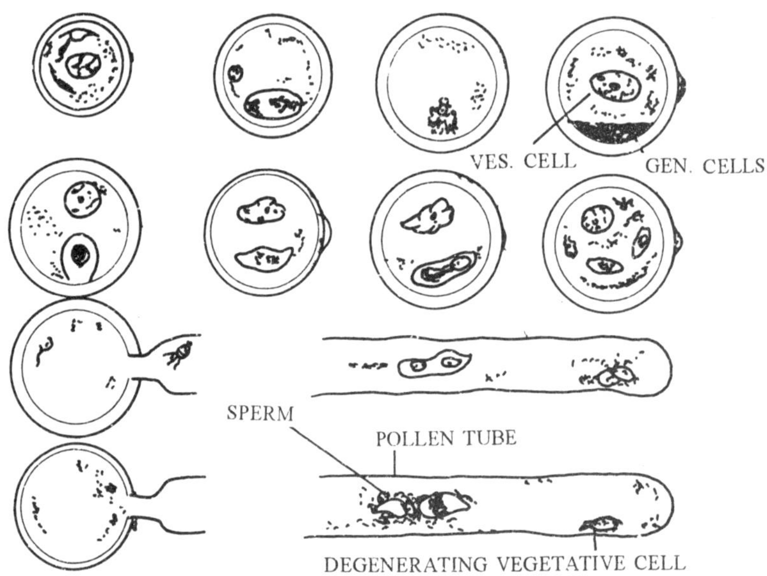


**Fig. 9.6** Stages in the development of Male Gametophyte

**Tapetum:** This is the innermost wall layer of the anther and is of considerable physiological significance because the food material to the sporogenous cells must pass through it. Typically the tapetum is single layered with cells having dense cytoplasm and a prominent single nucleus.

The tapetum is usually derived from the secondary pareital layer, though there are reports of its derivation from the sporogenous tissue (*Antirrhinum majus*). Based on the behaviour of the cells at the time of sporogenesis, the tapetum is divided into two types viz., *glandular* and *amoeboid.*

**Glandular** tapetum is also known as the *secretory type* of tapetum. In this type, the cells of tapetum remain unaltered throughout the development of the microspores. Nutritional materials are contributed to the anther sac by secretions from the inner faces of the cells until their total breakdown at the time of pollen maturity. In a great majority of Angiosperms (87%) the glandular tapetum is seen.

*Amoeboid* tapetum is also called *invasive* or *plasmodial* type of tapetum. Here, very early during microspore development, the tapetal cells breakdown, and the protoplasts 'wander' into the anther cavity encircling the pollen mother cells. The penetration of protoplasts may take place during meiotic prophase or may be delayed until the tetrad stage. This type of tapetum is seen in *Blainvillea, Emilia, Senecio* etc.

The close association of the tapetal cytoplasm with the microspores suggests that the former must be exerting a selective influence on the transport of material into the pollen.

**Sporogenous tissue:** The sporogenous cells may or may not divide to increase their number. Ultimately they round off and develop into *microspore mother cells* or *pollen mother cells* (PMC).

The PMCs undergo reduction division to produce tetrads of haploid microspores. At the initiation of meiosis, the PMCs get separated from the tapetal cells. Cytokinesis or wall formation takes place at the end of meiosis and is of two types. In the *Successive type*, a dyad is formed soon after meiosis. Wall formation is by a centrifugal plate of cellulose, which starts in the centre and extends laterally. The two cells of the dyad, divide next and four spores are formed. In the *Simultaneous type* the first meiotic division is not followed by wall formation, as a result a binucleate cells is formed. Both the nuclei divide and at the end of the meiosis cytokinesis takes place by centripetal furrowing.

As stated already, after meiotic divisions microspore tetrads are formed. These are of the following types viz., *tetrahedral* (three in one plane and one in the other), *isobilateral* (all the spores in one plane), *decussate* (spores arranged crosswise), *rhomboidal* (in the form of a rhomboid), *Linear* (spores are one above the other) and 'T' *shaped* (Two spores superposed and the other two juxtaposed).

Eventually the microspores separate from the tetrad and are set free in the anther loculus.

**Development of the male gametophyte:** Microspores represent the first stage of the male gametophyte. After their separation from the tetrad, microspores are called pollen grains.

To start with, the pollen grain has its single nucleus situated at the centre. The first division results in the formation of two unequal cells. The larger one is the *vegetative cell* and the smaller one is the *generative cell*. The vegetative cell eventually forms the pollen tube. The generative cell is attached to the wall of the pollen grain in the beginning, but later it comes to lie freely in the cytoplasm of the vegetative cell.

CHAPTER 10

# OVULE AND FEMALE GAMETOPHYTE

The ovule or megasporangium basically consists of two parts viz., *nucellus* and the *integument*. It is attached to the palcenta of the ovary by means of a short stalk called the funiculus. In a mature ovule the integument encloses the nucellus completely except for a small opening at one end called the *micropyle*. It is through this opening that the pollen tube gains entry into the ovule. The basal region of the ovule (region opposite the microphyle) is called the *chalaza*. Embedded in the nucellus is the female gametophyte or the embryo sac.

Mature ovules are classified into the following types, based principally on the position of the microphyle with reference to the funiulus.

**Orthotropous:** Here, the ovule is erect and the microphyle and the funiculus lie in a vertical line. This is seen in *Polygonaceae, Urticaceae, Piperaceae* etc.

**Anatropous:** The ovule takes a 180$^0$ turn so that the micropyle and the funiculus come to lie close to each other. This seen in almost all members of the sympetalae.

**Hemianatropous:** In this type, the ovule takes a 90$^0$ turn so that the microphyle is at right angles to the funiculus. This is seen in *Malpigiaceae, Primulaceae* etc.

**Campylotropous:** Here the ovule takes a turn a little less than 90$^0$ and is curved. This is seen in *Resedaceae, Leguminosae* etc.

**Amphitropous:** The curvature of the ovuie is so pronounced that it also affects the embryo sac which becomes horseshoe shaped. This is seen in *Alismaceae, Butomaceae* etc.

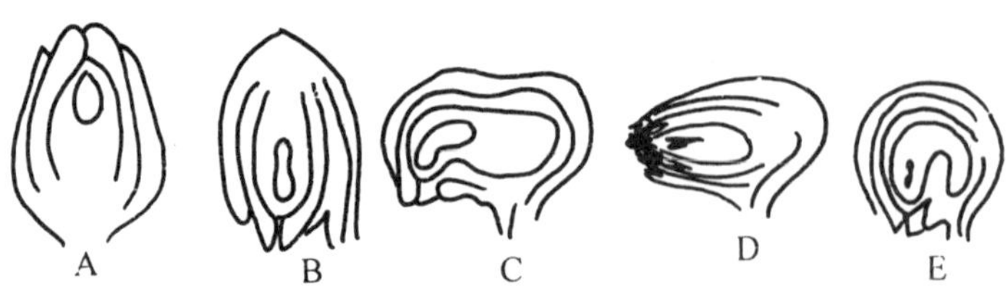


**Fig. 10.1** Types of Ovules A. Orthotropous, B. Anatropous, C. Hemianatropous, D. Amphitropous, E. Campylotropous

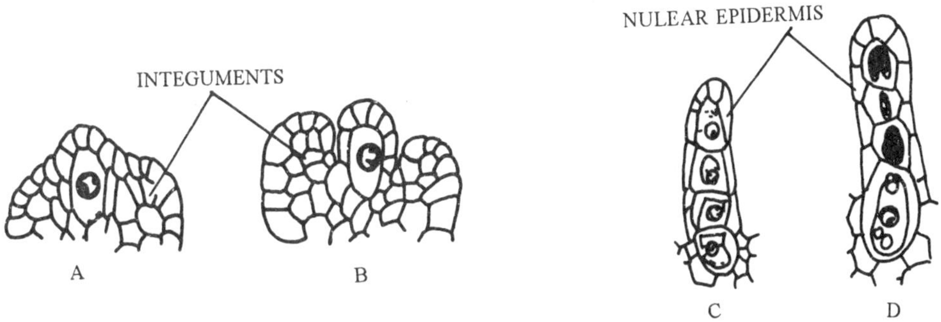


**Fig. 10.2 Female gametophyte**

**Megasporogenesis in a tenuinucellate ovule (redrawn from Johri and Singh, 1959)**

Circinotropous: Here the ovule first becomes anatropous. But subsequently the curvature once again lifts the micropyle to its original position so that the ovule takes a 360$^0$ turn. This type of ovule is seen *Plumbaginaceae* and *Cactaceae*.

**Integuments:** These constitute the envelopes of the ovule. In some cases as in *Loranthaceae*, the integuments are absent. Such ovules are said to be *atagmic*. Similarly ovules with one integument are called *unitegmic* and those with two, are called *bitegmic*. Rarely a third integument called *aril* is present in some members of *Ficoidaceae* (*Trianthema decandra*).

In bitegmic ovules, the micropyle may be formed only by the outer integument or only by the inner integument or by both the integuments.

**Nucellus:** Depending on the extent of the nucellus, the ovules are classified into two types viz., *tenuinucellate* and *crassinucellate*. In the former, the nucellus is scanty and the archesporium that arises in the hypodermal region directly functions as the megaspore mother cells (MMC), while in the latter the archesporium cuts off a few parietal cells and then functions as the MMC, so that the nucellus is bulky. A third category - *Pseudocrassinucellate* is proposed by Davis (1966), to include those ovules where the nucellar epidermis builds up the parietal tissue, while the archesporium directly functions as the MMC. But, here also the MMC becomes deep seated as in crassinucellate ovules.

## MEGASPOROGENESIS

A single hypodermal cell arises in the nucellus and it differentiates itself as the *archesporial cell* by its larghe size, conspicuous nucleus and dense cytoplasm.

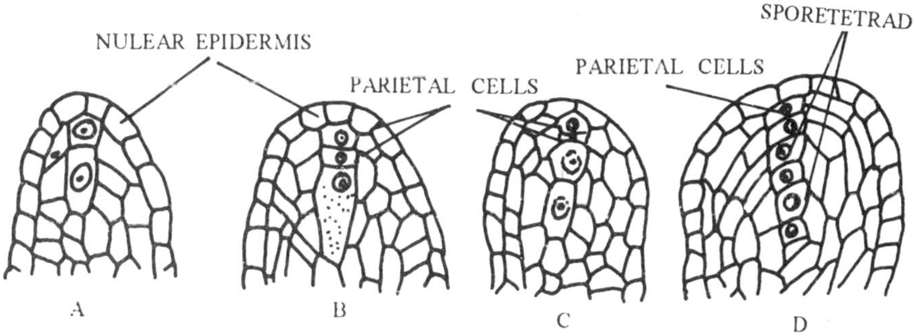


Fig. 10.3 Megasporogenesis

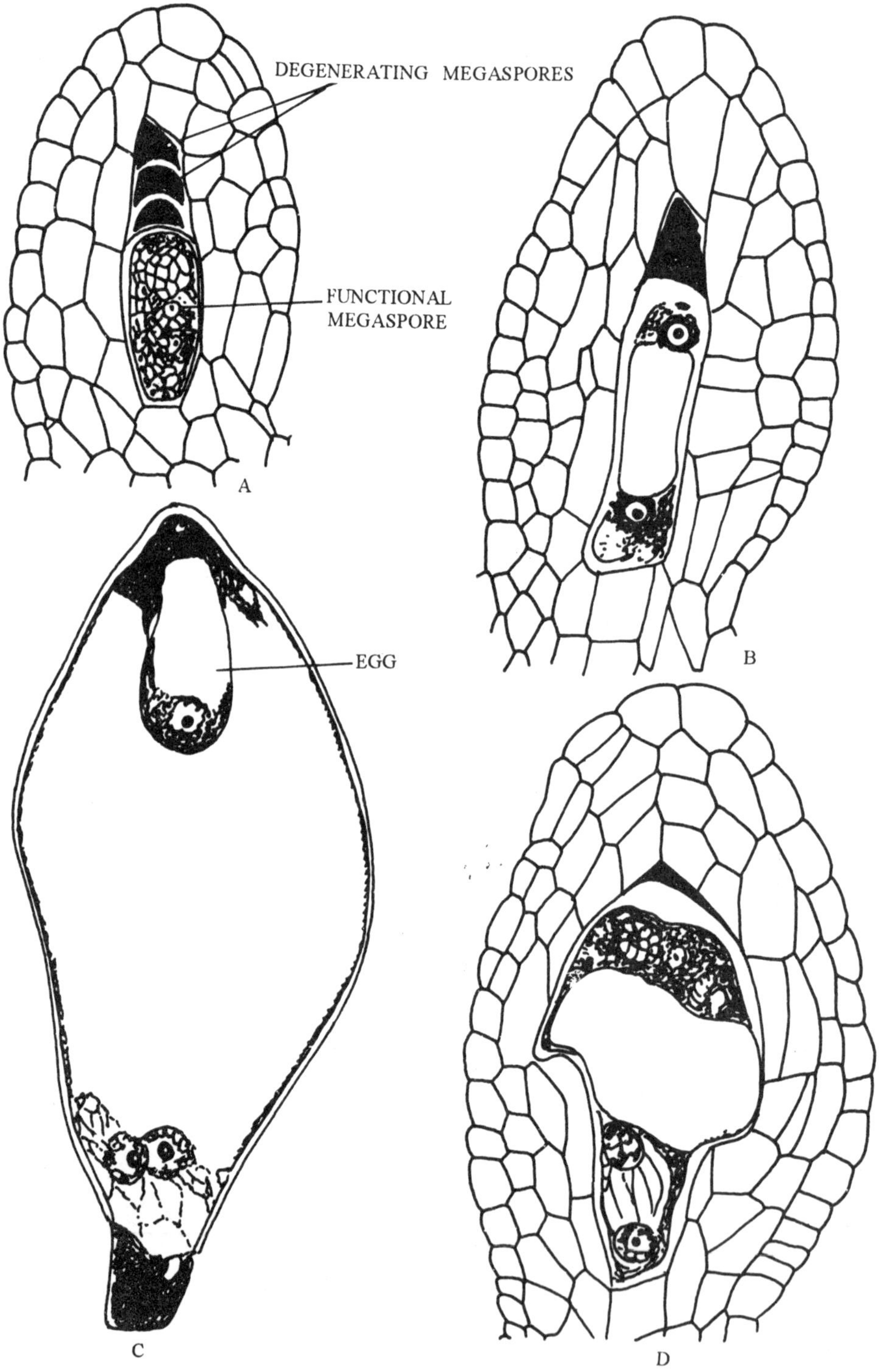


**Fig. 10.4 Female gametophyte**
Megasporogenesis in Hydrilla verticellata (redrawn from Maheshwari, 1950)

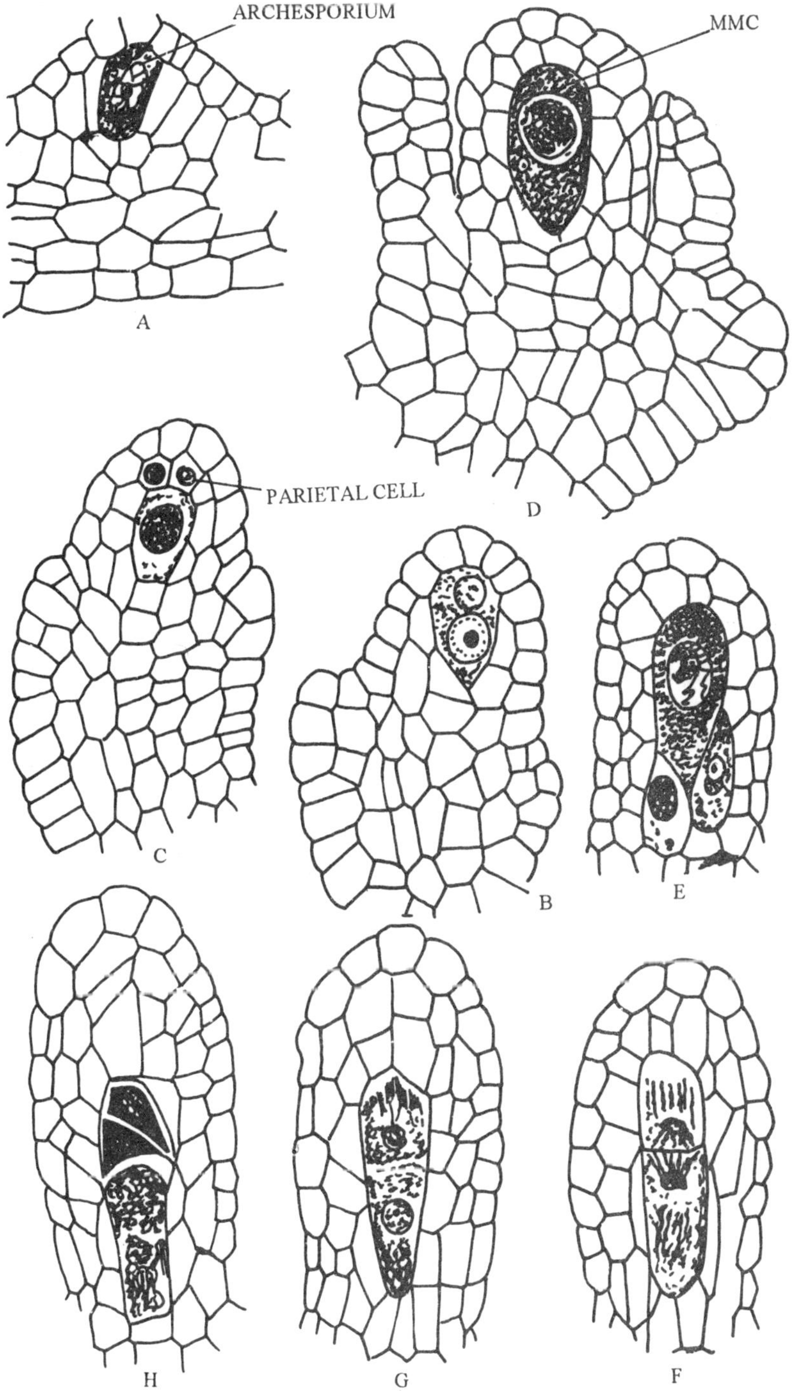


**Fig. 10.5 Female gametophyte**
Development of polygonum type of embryo sac in Hydrilla verticellata (redrawn from Maheshwari, 1950)

In tenuinucellate and pseudocrassinucellate ovules the archesporial cell directly functions as the MMC, while in crassinucellate forms it cuts off periclinally a parietal cell and then becomes the MMC.

The MMC undergoes reduction division. After the first meiotic division a dyad is formed by the laying down of a transverse wall. The second division is also transverses resulting in the formation of a linear tetrad of megaspores. Fequently the micropylar dyad divides vertically resulting in the formation of a 'T' shaped tetrad. Rarely the tetrad may be inverted T shaped (⊥), as in some Onagraceae. An isobilateral or tetrahedral tetrad of megaspores is very rare and it is reported only as an abnormality.

Of the four megaspores formed, usually the chalazal one is functional and it is called the functional megaspore. Occasionally the micropylar mehaspore may be functional. The three non functional megaspores degenerate and can be seen as black spots in a section of the ovule.

**DEVELOPMENT OF THE FEMAL GAMETOPHYTE**

The functional megaspore by further development forms the embryo sac (female gametophyte). Based on the number of megaspore nuclei that take part in the construction of the embryo sac, the embryo sac of Angiosperms is divided into three main types viz., *monosporic, bisporic* and *tetrasporic*. In the first one, only one spore gives rise to the embryo sac; in the second two spores and in the third all the four spores participate in the embryo sac development.

**Monosporic types:** The monosporic embryo sacs fall under two categories viz., **Polygonum type** and **Oenothera type**. The mature embryo sac of the former has eight nuclei while that of the later has only four nuclei.

**Polygonum type:** This is the commonest among flowering plants occuring in about 82% of the members. Hence it is often called the 'Normal type'. This was first discovered in *Polygonum divaricatum* (Strassburger, 1879). The MMC undergoes reduction division to form a degenerate and the lowermost (chalazal) one becomes funtional. This enlarges and its nucelus divides mitotically to form two nuclei. These two nuclei migrate to the opposite poles and a vacuole appears in between. This is the 2 - nucleate stage. The two nuclei divide again to form the four nucleate stage. All the four nuclei divide once again to form 8 nuclei (8 - nucleate stage) which are distributed at the opposite poles in two groups of four each.

After the 8-nucleate stage, there is no further division and the organisation of the embryo sac takes place. Of the four nuclei at the micropylar quartret, one migrates to the centre. Similarly from the chalazal quartret also one nucleus migrates to the centre of the embryo sac. The two nuclei fuse in the centre and form the diploid **secondary nucleus**. The three nuclei at the chalazal region form three uninucleate cells called the **antipodal cells**. The three nuclei at the micropylar region constitute the **egg apparatus**. The apparatus consists of two lateral **synergids** and a **median egg**.

In an organised polygonum type of embryo sac, 7 cells are present, namely, two synergids, one egge, a central cell consisting of the diploid secondary nucleus and three antpodal cells.

As stated already, there will be an egg apparatus, a secondary nucleus and three antipodal cell (except in *Oenothera* type) in the embryo sac.

**Egg apparatus:** The egg apparatus has two synergids and an egg. The synergids are elongated cells. They may be smooth or sometimes may be *hooked* or *beaked*. In some cases, the upper part of the cell is occupied by a *filiform apparatus*.

Electron microscopic studies have revealed that the filiform apparatus consists of a mass of inger like projictions of the wall into the cytoplasm. The lower region of the synergid has a large vacuole. The synergids are generally ephemeral, degenerating soon after fertilisation. But sometimes they may persist and become haustorial.

The egg is median and has a common wall with the synergids. The micropylar end of the cell has a large vacuole while the nucleus is situated at the lower end.

**Central cell:** This is the cell which consists of the secondary nucleus. Sometimes the two polar nuclei may not fuse for a long time and remain as free nuclei. It is from the secondary nucleus that the endosperm develops after fertilisation.

**Antipodals:** These are three in number and uninucleate. Sometimes they may be only two in which case one of them would be binucleate. The antiopodals generally degenerate soon after fertilisation. But frequently they may persist and become haustorial. Persistent antipodals are common in the tribe *Heliantheae* of *Asteraceae*.

CHAPTER 11

# ENDOSPERM

Endosperm is the chief nutritive tissue present in the embryo sac of Angiosperms. In Gymnosperms also there is an endosperm where the female gametophyte itself provides nutrition to the developing embryo. In Angiosperms however, the embryo sac is extremely reduced and by itself is incapable of supporting the embryo. Hence the need for endosperm arises. The origin of the Angiospermous endosperm is unique. It is formed by the fusion of three nuclei, of which two are female in origin and one is male. The resultant product i.e., primary endosperm nucleus (PEN) develops into the endosperm. Generally the endosperm is triploid, but the ploidy may vary as in *Oenothera* (diploid), *Peperomia* (polyploid) etc.

There are some exceptional Angiosperms where there is no endosperm formation. These are *Orchidaceae, Podostemaceae,* and *Trapaceae.*

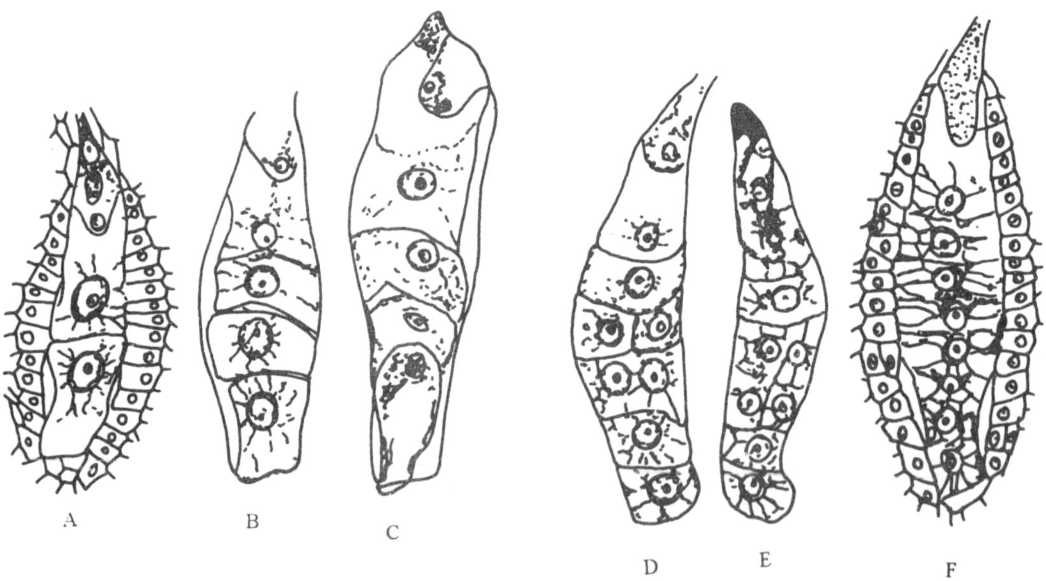


Fig.11.1 **Endosperm**

**Stages in the development of cellular endosperm in Villarsia reniformis (redrawn from Stolt,**

The endosperm produced may be scanty or bulky. In the former, (exalbuminous) in the mature seeds there will be no endosperm as it is completely consumed by the developing embryo. In the latter (albuminous) the mature seed has plenty of endosperm and it is the nutritive tissue in the seed.

**TYPES OF ENDOSPERM**

Based on the behaviour of the PEN, the endosperm is divided into three main types. These are *Nuclear, ab initio Cellular* and *Helobial*. Of these, nuclear endosperm is found in 161 families, while cellular is seen in 72 families and Helobial is seen in 17 families.

**Nuclear endosperm:** The PEN undergoes a few free nuclear divisions, without any wall formation. As a result, lying in the cavity of the central cell, a few to several thousand nuclei are formed. The free nuclear stage may remain until it is consumed by the embryo (*Limnathes*) or it may become cellular at a later stage. In the latter case cell wall formation takes place centripetally. The entire endosperm may become cellular or a part of it may become cellular. In *Phaseolous*, the endosperm, only around the embryo becomes cellular, while the rest of it is free nuclear.

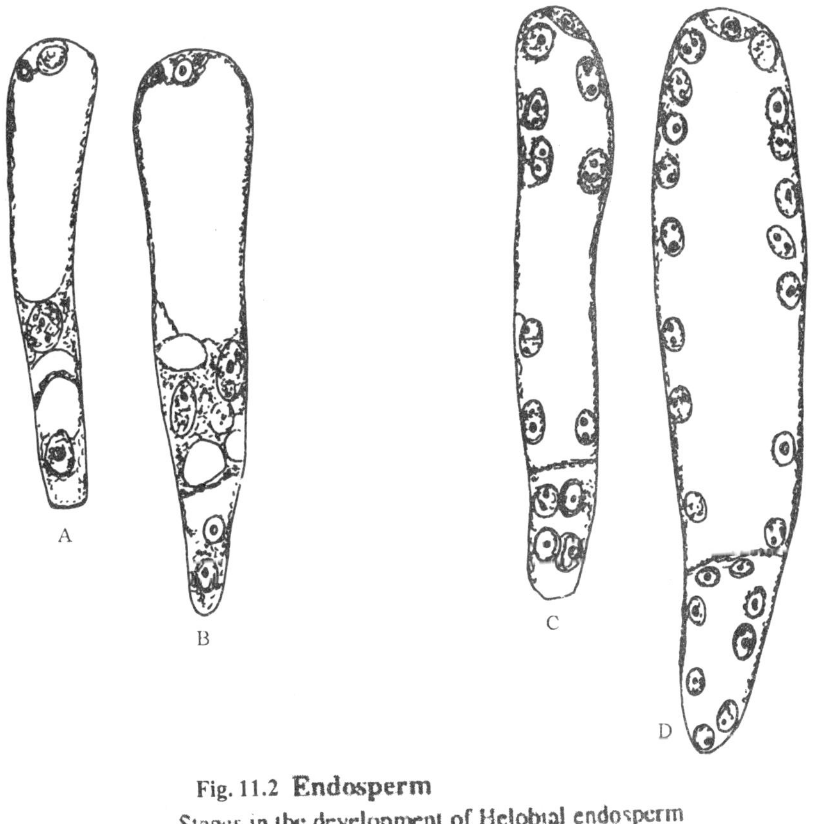


**Fig. 11.2 Endosperm**
**Stages in the development of Helobial endosperm**

**ab initio Cellular:** In this type, the first division of the PEN is always followed by wall formation, as a result two endosperm cells are formed. The first divisional wall may be parallel or perpendicular to the long axis of the embryo sac. Further divisions are also cellular and as a result the embryo sac gets filled by a cellular tissue.

Haustoria are very frequent in cellular endosperm also. These may arise either from the chalazal region or from the micropylar region or from both. Micropylar haustoria are seen in *Impatiens royeli* and *Hydrocera triflora* and many others. Chalazal haustoria are seen in *Magnolia obovata, Iodina rhombifolia* etc. In *Utricularia*, both chalazal and micropylar haustoria are seen.

**Helobial endosperm:** This is seen mostly in monocotyledons. This type of endosperm is intermediate between the nuclear and cellular types. The PEN migrates to the chalazal region of the embryo sac where it divides to form a large micropylar chamber and a small chalazal chamber. In the micropylar chamber, divisions are always free nuclear and cell wall formation, if any, starts very late. In the chalazal chamber, the nucleus remains undivided or divides only a few times. In some cases (*Phylidrum languginosum*), the chalazal chamber also becomes cellular.

CHAPTER 12

# EMBRYOGENY

## DICOTYLEDONOUS EMBRYO

There are about 5 types of embryo development found in dicotyledons. We will study a typical type viz., *Crucifer type*. Crucifer type of embryo is seen in the family *Cruciferae*. It is worked out in great deatail in *Capsella bursa-pastoris*. The foregoing description of embryogeny is based on the study of this plant.

The first division of the zygote is transverse and it results in the formation of a basal cell *cb* and a terminal cell *ca*. *ca* divides vertically while *cb* divides transversely to form a proembryonal tetrad which is 'T' shaped. Each of the two terminal cells (derivatives of *ca*) divide by a vertical wall at right angles to the first, resulting in a *quadrant*. The quadrant cells undergo a transverse division to form the *octant*. Of these cells, lower four give rise to the *stem tip* and *cotyledons*, while the upper four give rise to the *hypocotyl*. All these eight cells, next divide periclinally to cut off 8 outer cells which form the *dermatogen*, while the inner ones give rise to the initials of the *periblem* and *plerome*.

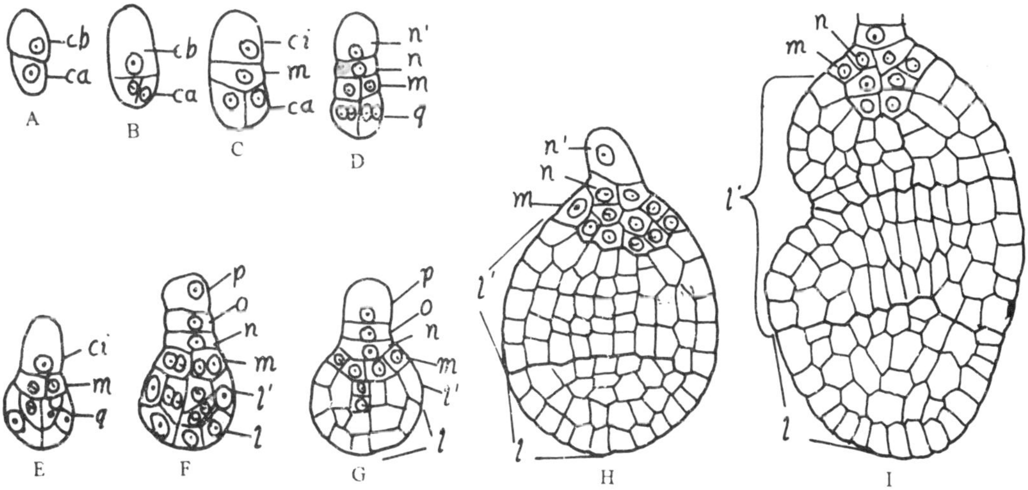


**Fig. 12.1 Embryo**

Development of embryo in Luzula forsten (redrawn from Soueges, 1923)

Meanwhile, the two deravatives of *cb* (produced by a transverse division) namely *ci* and *m* undergo a series of transverse divisions to form a row of 6-10 *suspensor cells*. The terminal most cell of the suspensor becomes large and vesicle like and probably serves as a *haustorium*. The lowermost cell of the suspensor (*h*) functions as the *hypophysis*. The hypophysis cell divides to form two tiers of four cells each. Of these, the lower tier forms the root cortex while the upper gives rise to the root cap and the root epidermis.

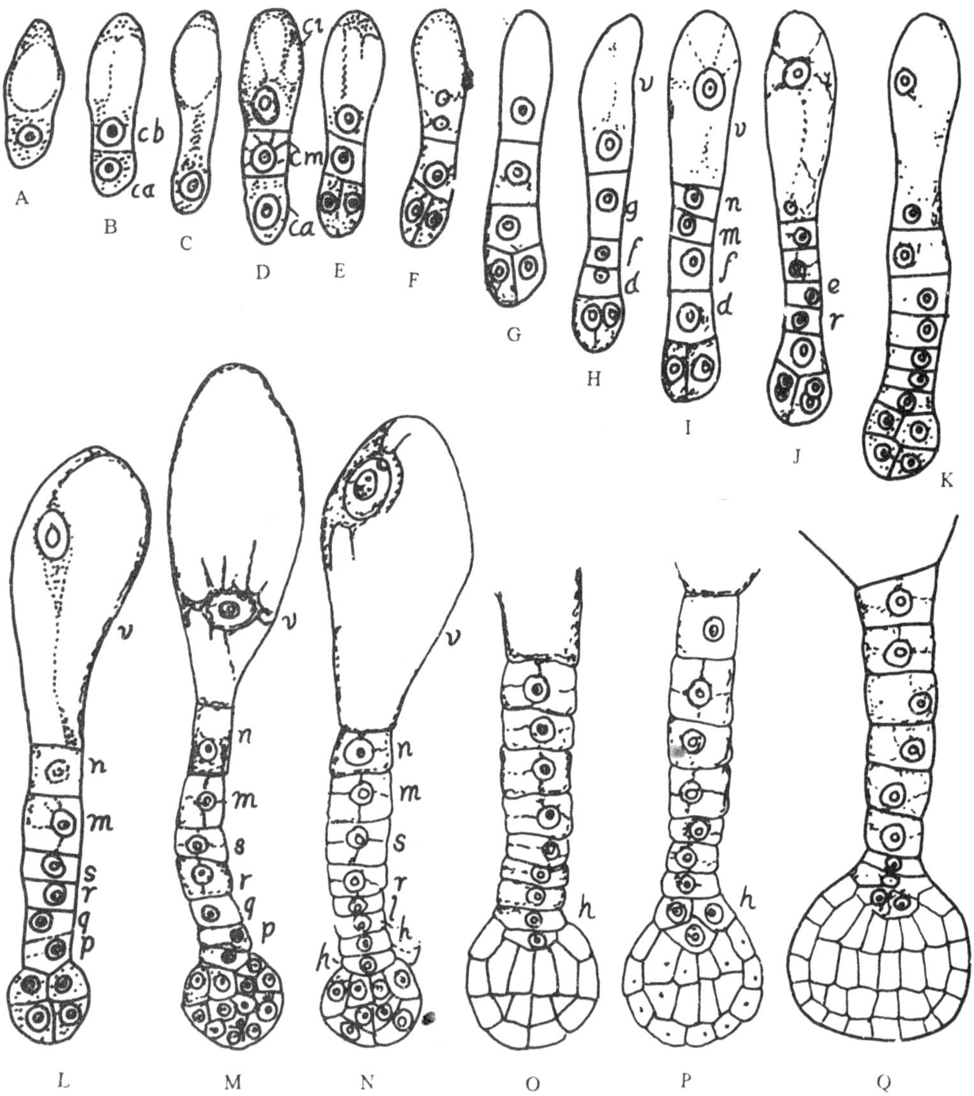


**Fig. 12.2 Embryo**

**Early stages of embryogeny in Capsella bursa - pastoris (redrawn from Soueges, 1914, 1919)**

At this stage, further growth in the embryo proper results in the formation of two cotyledons. The embryo now assumes a more or less cordate shape. The hypocotyl and cotyledons increase their size.

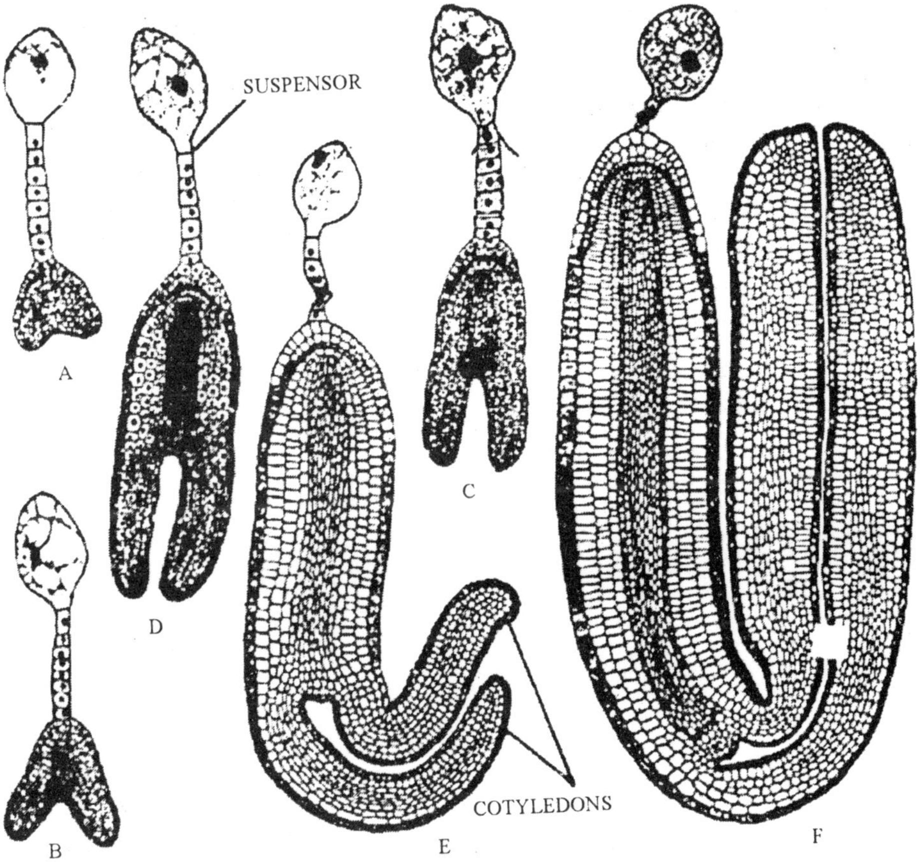


Fig. 12.3 **Embryo**

Later stages of embryogeny in Capsella bursa - pastoris (redrawn from Schaffner, 1906)

During subsequent growth the ovule becomes curved like a horse shoe and the cotyledons also are forced to assume this shape. In a mature embryo which is horse shoe shaped, the suspensor remains like a small tail.

**MONOCOTYLEDONOUS EMBRYO**

At stated already, there is not much of a difference in the development of embryo between monocotyledons and dictoyledons. In the mature embryo however, there is difference between the two. It is generally held that in a dicotyledonous embryo, stem tip is terminal and cotyedons are lateral, while in a monocotyledonous embryo it is *vice versa*. Some of the works of Swamy and his co-workers (Lakshmanan, 1972) however have established that even in monocotyledons, stem tip is terminal and later it is pushed towards a side due to rapid growth of the cotyledon. The developement of a typical monocotyledonous (*Luzula forsteri*) embryo proceeds as follows.

The zygote divides transversely to form a basal cell and a terminal cell. This is next followed by the formation of a T shaped proembryo. Quadrants are formed in the terminal cells by a vertical division at right angles to the previous plane of division. Periclinal divisions in the quadrant separate the dermatogen from the central cells. These cells divide transversely to form two tiers $l$ ,$l'$.

The tier *l*, gives rise to a part of the single cotyledon while the tier *l'* gives not only to the remaining part of the cotyledon but also to the hypocotyl and plumule.

While these developments are taking place, the basal cell divides transversely to form a row of three cells viz. *m*,*n*, and *n'*, *m* and *n* give rise to the root and root cap while *n'* gives rise to a single celled suspensor. In the mature embryo the cotyledon is terminal and the stem tip is *apparently* lateral.

# PART THREE
# ECOLOGY

CHAPTER 13

# ECOLOGY

Plants growing on the surface of the earth exhibit different characters under different environmental conditions. This is because, environment plays a profound role in the growth and development of plants. Plant also the animal life are affected by various ways. An intimate relationship exists betweeen organisms and their environment, so much so one can be regarded as theindex of the other. In short it can be safely said that plants and environment are interdependent.

Environment determines the density of plants and whether they can exist or not. The plants growing in any area fully adapt themselves to that habitat.

**What is Ecology?** The science of ecology also called environmental biology is extremely diversified and has transcended the limits of biology. In its broadest sense, Ecology deals with the analysis and study of inter - relationship between the organisms and their environment.

**History :** The term **ethology** a forerunner of ecology was first coined by a French Zoologist Isodore Geoffrey in 1859 to "study the relations of the organisms within the society and family in the aggregate and in the community".

Etymologically Ecology is derived from two Greek words *Oikos* (house) and *logos* (study), literally meaning the study of organisms in their dwelling place. The term Ecology was first used by Henry David Thoreau in 1858. According to some ecologists it was German biologist Reiter who should be credited with the coining of the word ecolcgy. It was another German biologist Ernst Haeckel (1870) who is supposed to have used the term in the sense we understand it today. He defined ecology "By ecology we mean the body of knowledge concerning the economy of nature - the investigation of the total relations of the animals both to its inorganic and organic environment, including above all, its friendly and inimical relation with those animals and plants with which it comes directly and indirectly into contact."

**Scope of Ecology :** Of all the branches of science, it is ecology that comes very close to human life, because human beings also interact with the environment and are influenced by it. Present day human problems are directly or indirectly related to ecology as their solution requires an understanding of Ecology. Ecology has been contributing very much these days to socioeconomic, political and other policies of theworld. The concern to conserve the environment for their own survival has awakened many people to rise against indiscriminate destruction of natural wealth.

The changes in climate, weather, monsoon etc are all related to changes in environment. A through knowledge of ecology is very much essential to monitor all these. Ecology thus plays a very vital role in human welfare in having its influence in many fields of human endeavour..

## STUDY OF VEGETATION

Vegetation study or study of the community structure is essential for understanding the dynamics of a plant community. Many of the features of a plant community such as density, species frequency etc are best expressed in terms of quantitative parameters. There are three methods of field study of the analytical characters. This are quadrat method, line transect method and point frame method. The following are some of the field experiments pertaining to the above.

***1. To determine the Minimum size of the quadrat by species Area-curve method.***

**Requirements** : Scale, four nails, A few meterres long string, hammer, graph sheet, herbarium sheet, cellotape etc.

**Procedure:** Hammer 3 nails to the ground in a 'L' shape and tie a piece of string. Now with another nail and a piece of string fashion a square of 10 x 10 cm. Make a note of the various species present in the area. This squarish area is called a quadrat. Increase the size of the quadrat to 20 x 20 $cm^2$ and note the species present. In the same way go on increasing the quadarat size to 30 x 30 $Cm^2$, 40 x 40 $Cm^2$ etc each time recording the number of species. Continue to increase the size of quadrat as long as there is increase in the number of species falling in the area. Stop increasing the size of the quadrat when there is no corresponding increase in the number of species.

Record the observations on a graph sheet. Size of the area to be marked on the X asis and the corresponding increase in the number of species on 'Y' axis. When you join the points on the sheet it would result in a curve. This is called the 'species area curve'. You will notice that initially there will be geometric increase in the number of species with the corresponding increase in the size of quadrat. Note the point where the curve starts flattening. Draw straight lines from this point to Y and X axis. The square area that will be obtained is the minimum quadrat size.

**Conclusion :** Species area curve is used to obtain the minimum size of quadrat while sampling the species composition of a plant community. Number of species (not individuals) in an area is a can be used as a parameter of diversity the number of quadrats. The curve obtained will indicate (where it starts to flatten) the point from which the minimum number of quadrats required can be found out.

***2. To determine the minimum number of quadrats to be laid down in the field.***

**Requirements :**Quadrat of known size, graph sheet, herbarium sheet, cellotape etc.

**Procedure :** Lay down about 30-50 quadrats of known size (as decided in the previous experiment) randomly at different locations in the area of study. List the various species found in each quadrat and find out the accumulation of species in each quadrat as shown in the table.

| Species | 1 | 2 | 3 | 4 | 5 | 6 | 7 | 8 | 9 | 10 |
|---|---|---|---|---|---|---|---|---|---|---|
| A | + | + | - | - | + | - | + | - | + | - |
| B | - | - | + | - | + | - | - | + | - | + |
| C | - | + | - | + | - | + | - | - | + | + |
| D | + | - | - | - | - | - | + | + | + | + |
| E | - | - | + | + | - | + | + | + | + | + |
| F | - | - | - | - | - | + | - | - | - | - |
| G | + | + | + | + | + | + | + | + | + | + |
| H | - | - | - | - | - | - | - | - | - | + |
| I | + | - | - | + | - | + | + | - | + | + |
| J | + | + | + | - | - | - | + | - | + | - |
| Accumulation of total No. of species | 5 | 4 | 4 | 4 | 4 | 5 | 6 | 4 | 7 | 7 |

Take a graph sheet and plot the accumulation of number of species (on the Y axis) against the numbe of quadrats. The curve obtained will indicate (where it starts to flatten) the point from which the minimum of quadrats required can be found out.

*3. To determine the density, frequency and abundance of different species in an area*

**Requirement :** Quadrat of known size, graph sheet, herbarium sheet, cellotape, pencil etc.

**Procedure :** Lay down a known number of quadrats (as determined in experiment 2) of definite size (as determined in experiment 1) in the area of study. List the various species present in each quadrat; also count the number of individual plants in each species (in every quadrat). Tabulate the data as shown in the following table.

| Name of species | No.of Quadrats land down 1 2 3 4 etc | Total No. of individuals in each species studied | No.of Quadrsts in which species species occur | Total number of quadrats | Comments |
|---|---|---|---|---|---|
| A | *2 *4 -- -- | 6 | 2 | 4 | |
| B | | | | | |
| C | | | | | |
| D | | | | | |

* No of individual plants occuring in that quadrat. From the data tabulated as above calculate the following.

**Density :** This is defined as the number of individual plants of a given species in a unit area.

$$\text{Density} = \frac{\text{Total number of individuals of the species}}{\text{Total number of quadrats studies.}}$$

$$\text{or} = \frac{\text{Toal number of quadrats in which the species occurs}}{\text{Total number of quadrats studied}} \times 100$$

**Frequency :** This is expressed in terms of percentage of distribution of species in a given area.

$$\text{Frequency (\%)} = \frac{\text{Total number of quadrats in which speices is present}}{\text{Total number of quatrats studied}} \times 100$$

Frequency (%) =

**Abundance :** This refers to the total number of individuals of a species in a given area.

$$\text{Abundance} = \frac{\text{Total number of individuals in all the quadrats}}{\text{Total number of quadrats in which the species has occured.}}$$

**Frequency Classes :** Frequency percentages can be arranged into the following five categories based on Raunkier's (1934) classification.

| Frequency Class | Frequency Percentage |
|---|---|
| A | 10-20 |
| B | 21-40 |
| C | 41-60 |
| D | 61-80 |
| E | 81-100 |

*4. To Study the vegetation by line transect method and to determine the frequency of individual speicies :*

**Requirements :** Measuring tape, graph sheet, herbarium sheet cellotape.

**Procedure :** Run the measuring tape to a suitable length ( 1 metre) and note down the species touching the tape. This length may be called one unit of a line transect. Run several such line transects randomly in different sites in the aea of study and note down the number of species . In addition count the number of individuals of each species touching the tape in each unit of the line transect. Tabulate the data as shown in the following table.

| Name of species | Units line transect<br>1 2 3 4 etc | Total No. of transects in which the species present | Total No. of line transects studied | Frequency Percentage | Frequency class |
|---|---|---|---|---|---|
| A | * + + + +<br>(3) (2)(4) (1) | | | | |
| B | - + + -<br>(3) (2) | | | | |
| C | - + - +<br>(1) (2) | | | | |
| D<br>etc. | | | | | |

* + indicates presence of species (not number of indivudals)
- indicates absence of species

** Value in parenthesis shows the number of individuals of particular species (A, B etc) in a given transect.

Calculate the frequency percentage for each species as in quadrat method.

Density of each species also may be calculated as in quadrat method.

*5. To study the vegetation by point frame method and to determine the frequency of individual speicies.*

**Requirements :** Point frame, graph sheet, herbarium sheet, cellotape etc.

**Procedure :** The point frame apparatus consists of a wooden frame (see figure) in the form of a rectangular bracket. The long horizontal arm has a number of holes to accomodate pins (nails).

Place the point frame apparatus (with 10 pins) in the field. The area occupied by the apparatus constitutes one sampling unit. Place the apparatus randomly in several places in the field and note down the number of species touching the pins. Also note down the number of individuals in each species touching the pins. Tabulate the data as in the case of transect method (substitute one unit point frame for one unit of line transect) and calculate the frequency percentage, frequency class and density.

*6. To study the vegetation by Biological Spectrum Method*

**Requirements :** Scalpel, scissors, graph sheet, spatula graph sheet, herbarium sheet, cellotape etc.

**Procedure :** In a given area of study examine each species as to the location of pernneating buds (or seeds) on the plant body. Note down the number of species. Now categorize the various species into the following life form types based on RaunKaier's classification as follows

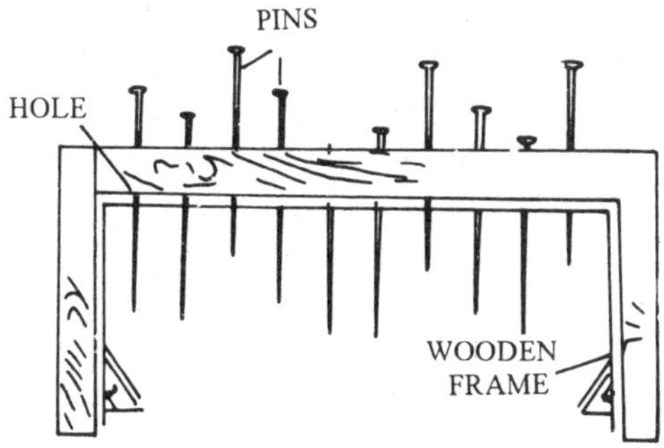


**Fig 13.1 Point frame apparatus**

1. Phanerophytes - Trees, shrubs with buds atleast 25 cms above ground.
2. Chamaephytes - Prostrate plants with buds on the surface or upto 25 cm above ground
3. Hemicryptophytes - Perennial plants with buds above or little below the soil surface.
4. Geophytes - Plants Rhizomatous or have under ground buds
5. Therophytes - Seasonal plants

Calculate the percentage value for each life form against the total life form in the area. With the help of a graph sheet plot the various life forms on the X axis and their percentage on the Y axis. Either a curve or histograms can be drawn on the graph sheet. This will provide the biological spectrum of the area.

Biological Spectrum may be defined as the distribution percentage of various life forms in a given area :

7. ***To determine the biomass of green plants in a given area***

**Requirements :** Quadrat of known size, polythene bags, spatula, hot air oven etc.

**Procedure :** Lay down a number of quadrats randomly in a given area. Collect the plants in the quadrats as follows :

1. Cut all the aerial parts and put them in a polythene bag.
2. Collect all the dead, fallen leaves, twigs etc and put them in another bag.
3. Dig out all the underground parts of plants clean them of soil and collect in another bag.

Weigh each portion (1, 2, 3) separately and note down the fresh weight. Now put them separately in an oven at 90°C for 24 hours and weigh them again. Note down the dry weight.

a. Fresh weight of all plant parts $X_{gms}$
b. Weight after (dry weight) $X_{1gms}$
C. Moisture content $X_{gm}$ - $X_{1gm}$

Bio mass per quadrat (unit area) may be caluclated as follows. Suppose the quadrat is 50 x 50 $cm^2$ - 2500 $cm^2$. This means 2500 $cm^2$ area has $X_{1gms}$ of dry matter. Therefore one square metre will have

$$\frac{X1 \times 100 \times 100}{2500} = X_{2gm}$$

The biomass of plant parts is $X_{2gms}$ per square metre

7. ***To determine the vegetational cover in an area***

**Requirements :** Measuring tape, scissors, Vernier Calliperse, paper etc.

**Procedure :** Run the measuring tape for specific lengths at several places. Note down the length of plants touched by the tape.

Cut a transverse section of some plants at the ground surface. Measure the diametre of the cut end with the help of vernier scale. Tabulate the data as follows.

| Name of Species | Units of tape<br>1 2 3 4 5 etc | Total |
|---|---|---|
| A<br>B<br>C<br>etc. | | |
| Total Number of Species | Length of tape for individual Plants | Total Length of all individuals |

From the data above calculate the total length of tape (number of units x length of each unit). Covered by all the species and percentage of of total length of tape covered by each species. This gives the % vegetational cover of the area.

| Name of species | Diametre<br>No. of Individuals<br>1, 2, 3, 4 etc. | Total | Average |
|---|---|---|---|
| A<br>B<br>C<br>etc | | | |

Calculate the basal area as follows

basal area = $\pi r^2$ where

$r$ = radius

$\pi$ = 3.14

Average basal area for each species may be calculated by totalling the basal area of all species and dividing it by the number of species.

*8. Study of Soil texture :*

**Requirement :** Polythene bag, pick axe, field lens, forceps, mesh of different sizes of pores.

**Procedure :** Collect soil samples from various sites in polythene bags. Examine the dry sample by a field lens. Examine a moist sample also. Squeeze the soil sample under thumb and feel the texture.

**Categorize the soil as follows :**

| | | |
|---|---|---|
| 1. | Particles very large | Gravel |
| 2. | Individual grains may be seen and felt between fingers. Do not adhere when | wet Sand |
| 3. | Individual grains are seen and felt, form a cast when moist but fall apart when squeezed | Sandy loam |
| 4. | Dry soil mellow with a gritty feel. Moist soil forms a cast which does not break while handling. | Loam |
| 5. | Dry, appears cloudy, soft feeling, forms a cast (both dry and wet) of plastic material when wet | Silt loam |
| 6. | Hard when dry but forms cast which does not break while handling. Moist soil when pressed forms thin ribbon | Clay loam |
| 7. | Very hard lumps when dry, sticks to surface when moist; forms long ribbon when squeezed | Clay |

By using meshes of different pore size also soil particles can be classified.

| Diameter of the Particle in mm | Classification of particle |
|---|---|
| Less than 0.002 | Clay |
| 0.002-0.02 | Silt |
| 0.02-0.2 | Fine sand |
| 0.2-2.0 | Coarse sand |
| 2.0-5.0 | Fine gravel |
| More than 5.0 | Coarse gravel |

*9. Study of soil temperature*

**Requirements :** Soil thermometer (ordinary thermometer also can be used) Iron rod etc.

**Procedure :** (i) Dig a hold of desired depth with the help of an iron rod and directly measure the temperature by inserting an ordinary thermometer. Record the reading after about 15 minutes.

(ii) Soil thermometers which have a steel end near the mercury bulb can be directly pushed into the soil to the desired depth and temperature measured.

***10. Measurement of soil pH***

**Requreimennts :** Soil samples, water, barium sulphate, Universal indicator or pH paper.

**Procedure :** Collect different samples of soil in separate test tubes. To each test tube add about 5 ml of barium sulphate and 15 ml of water. Keep the solution aside untill a clear supernatant fluid is formed. Take few drops of this on a porcelain tile and add few drops of universal indicator. Note the change in colour and compare it with the colour chart for different pH values pasted on the bottle. Alternatively dip a piece of pH paper in the fluid, note the change in colour and compare it with the colour chart.

***11. To determine the moisture content (%) in soil***

**Requirements :** Soil samples, hot air oven, china balance etc.

**Procedure :** Place soil samples of known weight in a hot air oven at a temperature of 105-110°C for 24 hours. Weight the soil sample before keeping them in hot air oven and also weigh after heating . Calculate the moisture percentage as follows.

1. Weight of soil before heating - $X_{gms}$
2. Weight of soil after heating for 24 hrs $X_{1gms}$
3. Moisture content $X_1$ - X
4. % moisture content = $\frac{X-X_1}{X_1} \times 100$

***12. Water holding capacity of soil***

**Requirements :** Soil samples, water, seive, oven, dessicator, balance etc.

**Procedure :** Pick a number of small holes in the base of the tin box. Fill the box with air dried and seived soil. Add water so that the soil gets saturated. Keep the tin in a slanting position and hang it to a stand with the help of string. Extra water comes out of the perforations at the base. Allow it to stand for a while. When water drops stop coming out remove the soil and weigh immediately. Note down the weight. Afterwards keep the soil in a hot air oven and dry it at 105°C for 48 hrs. Cool the soil sample in a dessicator and record the weight.

**Observations and Conclusion :** The difference in the weight in soil samples before and after heating gives the amount of water in the soil or water holding capacity of the soil. Calculate the water holding capacity per gram of soil. Also calculate the percentage of water retained by the soil. This is generally refered to as the field capacity of the soil.

***13. Calculation of carbonate content, nitrate content and base dificiency of soil samples.***

**Requirements :** Soil samples HCL, diphenylamine, $H_2SO_4$ ammonium thiocyanate, $H_2O_2$, water, china clay, tiles, test tubes etc.

**Procedure :**

(a) **Carbonate :** Take a little amount of soil sample in separate test tubes. Add a few drops of Hcl to each of the tubes. Observe the rate of effervescence. Compare the rates in different soil samples. Higher rate of effervscence indicates higher carbonate content.

(b) **Nitrate content :** Prepare a soil solutions of different samples by adding 5 parts water to one part water. Take a few drops of soil solution on a porcelain plate. Add a few drops of diphenylamine and a few drops Conc $H_2So_4$. Blue colour formation indicates the presence of nitrates. Higher intensity of colour indicates higher nitrate content. Different soil samples may be compared based on the intensity of colour.

(c) **Base defeciency :** Take a pinch of soil sample in a tube, add saturated alcholic solution of ammonium thiocynate and shake vigourously. Allow the contents to settle. Coolect the supernatant and to it add a few drops of $H_2O_2$ and observe the intensity of red colouration that develops. Compare the degree of colouration in different soil samples. Higher intensity of colour indicates increased dificiency of replacable bases.

### *14. Test for the presence of Inorganic Salts*

**Requirements :** Soil samples, distilled water, barium chloride solution, Hcl, $H_2So_4$, Nitric acid ammonium nitrate, ammonium molybdate, silver nitrate, beakers, test tubes, conical flask etc.

**Procedure :** Take 200 gms of soil sample in a conical flask and add 500 ml of water. Shake vigourously and keep it aside overnight to allow soluble salts to dissolve; filter the water and collect the filterate (aqueous soil extract) conduct the following tests.

(a) **Chloride :** Take 10 ml of aqueous soil extract in a test tube and add 10 ml of 0.1 N $H_2So_4$ to neutralize the carbonate and bicarbonate content. Now add silver nitrate to the mixture. Development of white precipitate indicates the presence of chloride.

(b) **Phosphate :** Take 10 ml of aqueous soil extract. Add a few drops of ammonium molybdate solution, Conc. nitric acid and ammonium nitrate. Development of yellow colour indicates the presence of phosphates.

(c) **Sulphate :** Take 10 ml of aqueous soil extract in a test tube and add 5 ml of Conc. HCl and boil cool and add a few ml of $BaCl_2$ solution. Development of white precipitate indicates the presence of sulphate.

### *15. Measurement of temperature, relative humidity and light intensity of air (atmoshphere).*

**Requirement :** Maximum - minimum therometer, and wet thermometer (psychrometer) photometer etc.

**Procedure :** Temperature of the atmosphere can be measured by means of 'U' shaped thermometer in which one arm records the minimum and the other maximum temperature. The 'U' Shaped thermometer may be kept in different places like open air, shade etc and the temperature may be recorded.

**Relative Humidity :** This is recorded with the help of psychlometer also know as dry and wet thermometer (see fig). THe psychlometer has two thermometers - one dry and the other wet. The bulb of wet thermometer is always kept moist by dipping it in moist cotton. Evaporation of water due to normal temperature of the atmosphere causes cooling of the bulb of wet thermometer and lowers the temperature. Dry thermometer records the normal temperature. Note the difference in temperature between dry and wet thermometers. Relatiive humidity may be calculated by using a conversion table. If the relative humidity is 100% there will be no difference in temperature between the dry and wet thermometer.

### *16. Measurement of light intensity*

**Requirements :** Photometer (Lux meter)

**Procedure :** Photometer consists of a photo electric cell and an amperemeter (microammeter). The instrument has a filter to guard against intense light. Expose the apparatus and read the value with the deflection of the needle. The photo electric cell on exposure to light converts it into electricity and the current generated is measured in the ammeter in units called lux units.

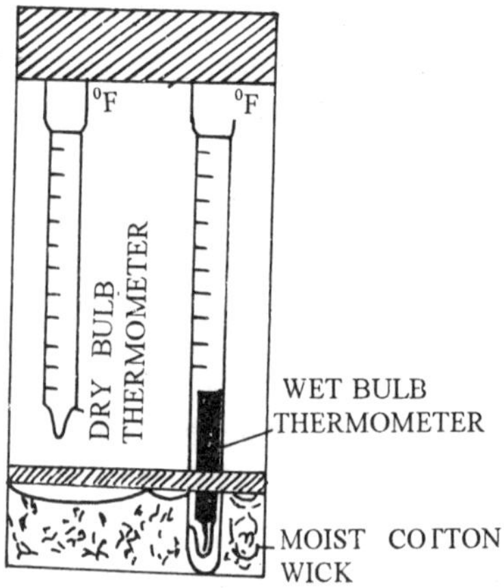


Fig 13.2 Dry and Wet Thermometer

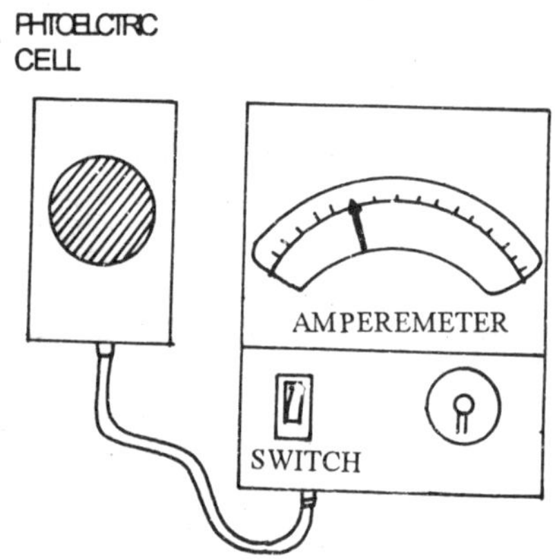


Fig 13.3 Light Meter

*16. Measurement of turbidity of water :*

**Requirements :** Secchi disc

**Procedure :** Secchi disc is a circular piece of metal having black and white strips. The disc has a hole in the centre through which a thread tied to a weight passes (through). The disc is lowered to a level where it will not be possible to distinguish black and white strips. If the depth at which the black and white strips lose their identity is more then the turbidity is less and vice versa. This can be compared in different ponds or tanks.

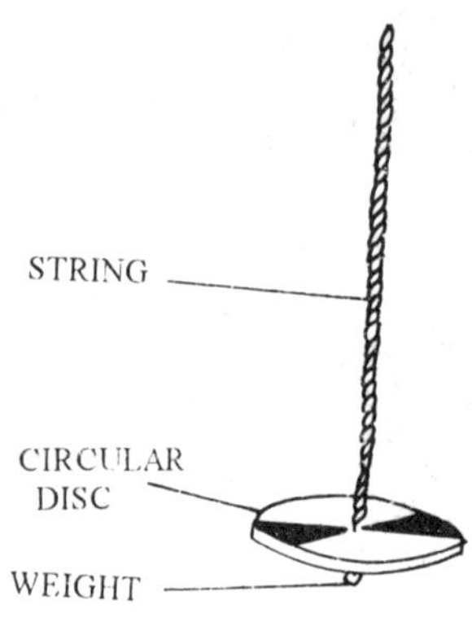


Fig 13.4 Sacchi Disc

*17. Measurement of rainfall of a given area :*

**Requirement :** Rain guage (measuring cylinder, funnel, jar, cup, oil etc).

**Procedure :** Keep the measuring cylinder in a jar and pour a little oil into the cylinder so that it forms approximately a 5mm thick layer. This prevents surface evaporation. Place a funnel over the measuring cylinder and fix a cap over the jar. The cap allows water only into the funnel and not into the jar. Place the apparatus in an open area and watch.

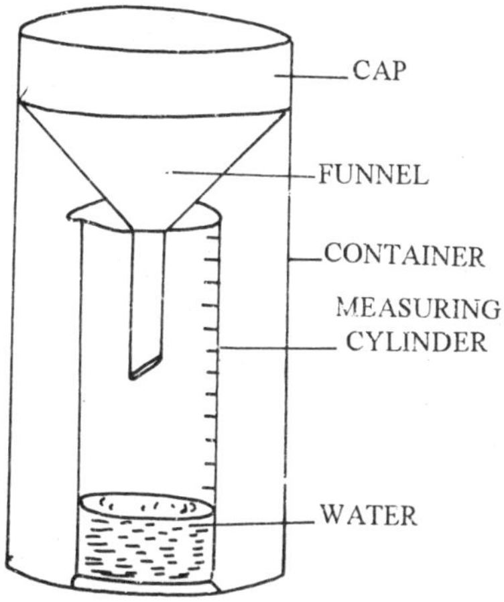


Fig 13.5 Rain gauge

**Observation and Inference :** Rain water percolates through the funnel and collects in the measuring cylinder. Measure the diameter of the cap and the amount of water collected. Calculate the amount of rainfall per unit area.

## *ECOLOGICAL GROUPS*

Groups of plants belonging to one or more species that grow together in a particular area constitute an ecological community or a group.

For instance trees shrubs, climbers etc that grow in a forest form a forest community. Similarly plants that grow in a pond form a pond community. Likewise, plants that grow in desert form a desert community. Based on the environment they inhabit warming (1985) has identified eight ecological groups or communities.

## HYDROPHYTES

Plants that grow in water or in soil covered with water belong to this group. Hydrophytes have a number of morphological and anatomical adaptations. These are

### Morphological Adaptations :

1. **Dimorphic leaves :** Some of the plants which are partially submerged have dimorphic leaves. The underwater leaves are thin and highly dissected whereas the leaves above water are broad. This condition is known as heterophylly. Ex. *Limnophylla heterophylla, Trapa, Sagittaria sagittifolia* etc

2. **Formation of coiled pedicles and petioles :** In plants like *Vallisnaria* the pedicles and petioles are coiled so that the flowers and leaves will be always on the surface of water but do not ofter resistannce to water currents.

3. **Development of water Proofing device :** To prevent decay a waxy coating or mucilaze covering is developed on the surface to repel water.

4. **Roots :** Highly reduced, divided into thin fibrels or even absent (ultricularia). Root hairs and root caps are also absent very often.

5. **Stem :** Very thin, delicate and transparent.

6. **Guttation :** Many of the plants remove water by way of guttation.

### Anatomical Adaptations

1. **Epidermis :** Cells are thin walled, contain chloroplasts

2. **Aerenchyma :** Parenchyma tissue has a large number of air cavities to store air which helps both in respiration and buoyancy (floating).

3. **Cuticle :** Absent or even when present very thin

4. **Mechanical Tissues :** These are very poorly developed. Sclerenchyma and xylem which are the principal lignified tissues are very little in extent.

5. **Stomata :** Absent in submerged leaves, but present in the upper epidermis of floating leaves

6. **Secondary growth :** Takes place very rarely.

7. **Chlorenchyma :** Very well developed.

**Classification of Hydrophytes**

1. **Floating hydrophytes :** Plants float on the surface of water. There are two kinds of surface hydrophytes - free floating and fixed. In the former the roots are suspended freely in water (e.g. *Eichhornia, Pistia, Salivnia* etc) Fixed floating hydrophytes have their roots fixed to the underwater soil. They occur in shallow water (e.g *Marsilea, Nymphaea* etc).

2. **Submerged hydrophytes :** They are completely under water; may be suspended freely in water (e.g. *Elodea*) or may be rooted to soil underneath the water (e.g *Vallisnaria*).

3. **Amphibious Hydrophytes :** They grow in shallow or muddy waters. Roots and parts of their stems are embedded in water while their branches and leaves are often aerial (e.g. Typha, Alisma etc). We will now study in detail morphological and anatomical adaptations in a few exapmles.

**Hydrilla verticellata (Hydrocharitaceas)**

Fig 13.6 Hydrilla - Habit

**Morphology**

1. It is a commonly occuring water plant in ponds.
2. It is a submerged hydrophyte, but rooted to the under water soil.
3. Root system is adventitious.
4. Stem is transparent, flexible and soft.
5. Leaves are small, thin and transparent. Phyllotaxy is whorsed.
6. Mucilage covers all parts of the plant body to prevent excessive absorption.

**Transverse Section of stem**

1. Epidermis is single layered made up of isodiametric thin walled cells.
2. Cuticle if any is very thin
3. These is a single layer of parenchymatous hypodermis
4. Cortex is made up of aerenchyma. Air chambers are irregular in size and shape and are separated. by one layered partititon of cells. Cortical cells are spherical or oval and thin walled. Air cavities are larger in size towards periphery but smaller towards the centre.

5. Internal to the aerenchymatous cortex is the endodermis surrounding the stele.
6. The stele is highly reduced. Xylem is absent and is replaced by an air cavity. Surrounding the air cavity is phloem.

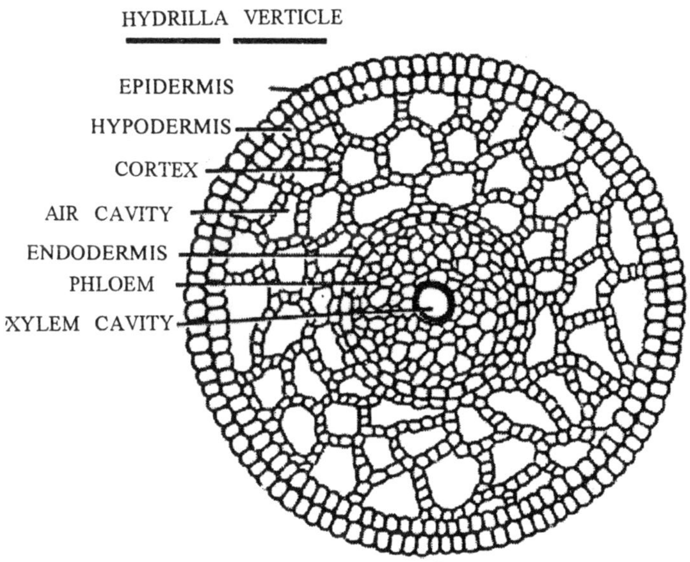


**Fig 13.7 Hydrilla, T.S. of Stem**

### Transverse Section of Leaf

1. Leaf is dorsiventrally flattened and reduced in size.
2. The upper and lower epidermal layers are made up of a single layer of cells each. There is no cuticle but there is deposition of mucilage.
3. There are no stomata
4. Vasculature is highly reduced. There is no bundle sheath
5. Mesophyll is also highly reduced in quantity and undiffernTiated
6. All cells including epidermis are chlorophyllous

### Nymphaea (Nymphaeaceae)

### Morphology

1. It occurs in shallow waters
2. Plant is a submerged rooted hydrophyte
3. Stem is rhizomatous growing horizontally underneath the soil.
4. Roots are adventitious.
5. Leaves are very large, and have a long flexible petiole.

6. Lamina is cordate is shape and has waxy coating as a result water does not adhere to the surface.

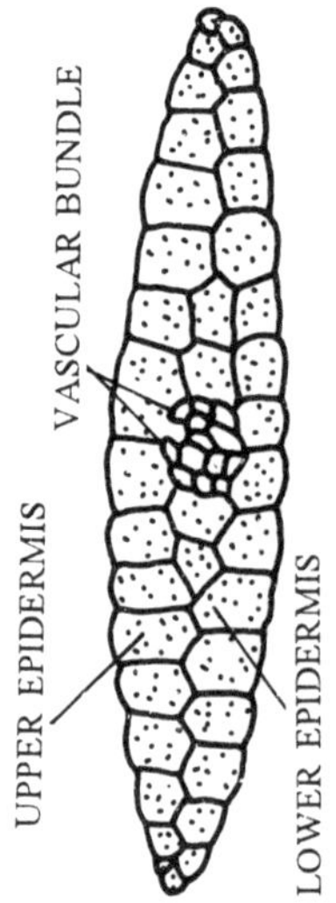


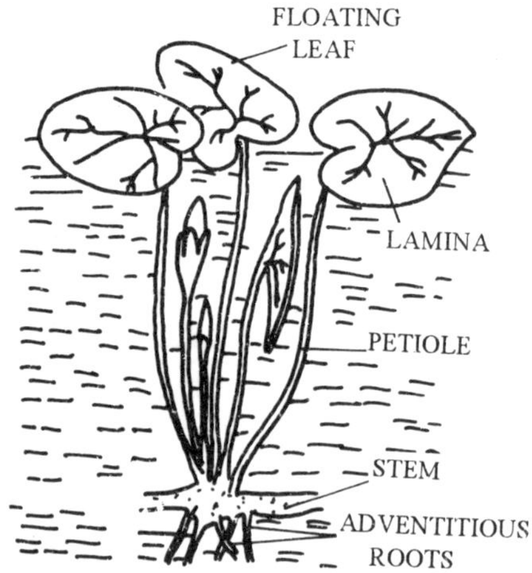


**Fig 13.9. Nymphaea, Habit.**

**Fig 13.8. Hydrilla, T.S. of Stem**

**Transverse Section of Petiole**

1. Outermost layer is epidermis. It is made up of thin walled barrel shaped cells. Cells are compactly arranged. There are a number of multicellular hairs arising from the epidermis. A thick coating of mucilage is found on the epidermal walls.
2. Internal to the epidermis is a few layers of collenchyma constituting the chief mechanical tissue. (Collenchyma is a living mechanical tissue whose corner walls are thickened). The collonchyma constitutes the hypodermis.
3. Next to the hypodermis is the ground tissue. Immediately next to hypodermis there are a few layers of compactly arranged cells (parenchyma) with small intercellular cavities. Next to this is the aerenchyma with large air spaces separated by uniseriate partitions.
4. Branched or stellate sclerids are found protuding into the air cavities.
5. Vascular bundles are reduced. They are scattered in the ground tissue being present mostly in the outer zone.
6. In the vascular bundle xylem is represented by a lacuna. Phloem elements are present in their normal condition.

**Transverse Section of leaf (lamina)**

1. The lamina is dorsiventrally flattered (bifacial)
2. Upper epidermis is single layered. waxy cuticle is deposited above the epidermis.
3. Stomata are found only in the upper epidermis (epistomatic).

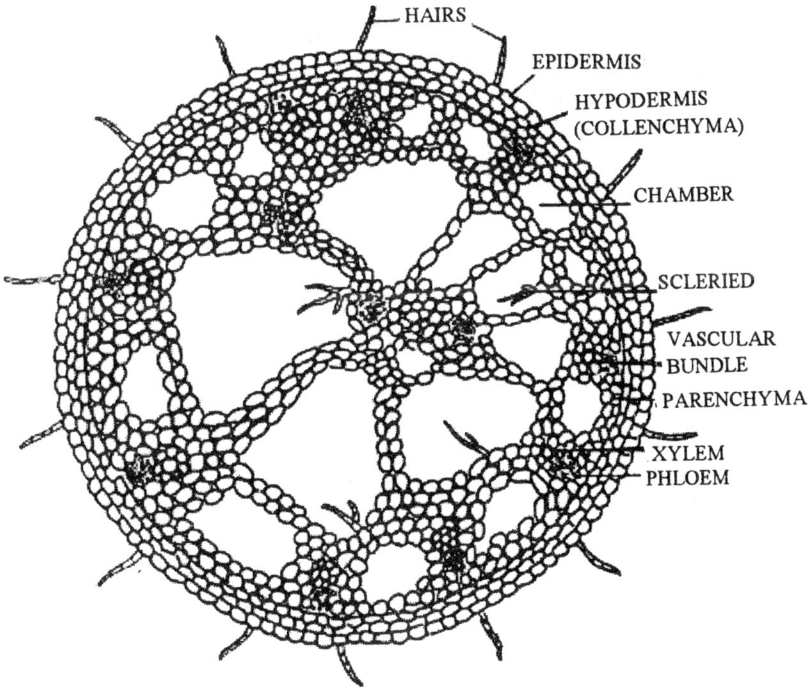


**Fig 13.10 Nymphaca, T.S. of Petiole**

4. Lower epidermis is also made up of a single layer of cells. There is no cuticle on lower epidermis.
5. Mesophyll is differntiated into upper palisade parenchyma and lower spongy parenchyma.
6. Palisade parenchyma is 2 layered. Cells are compact. This is the chief phtosynthetic tissue in the leaf.
7. Spongy parenchyma cells have very large air cavities between them.
8. Macro sclerids (in palisade tissue) and a sterosclerids (in spongy tissue) are present in the mesophyll.
9. Vascular bundles are poorly developed. They contain protoxylem cavities. Phloem is well developed.

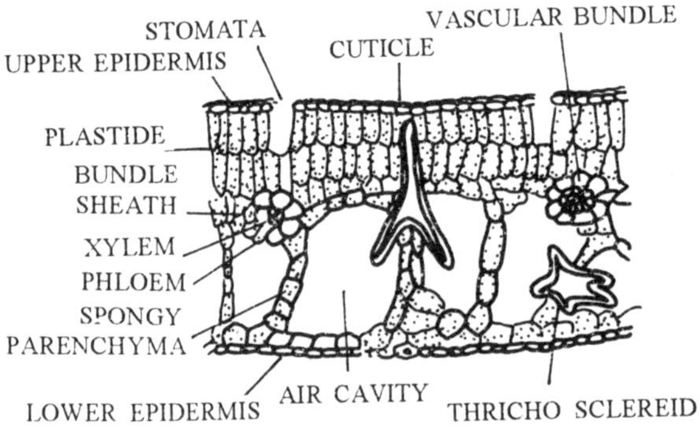


**Fig 13.11 Nymphaea, T.S. of Lamina**

**Transverse Section of Stem**

1. There are three regions in a transverse section epidermis, cortex and stele.
2. Epidermis is single layered. Multi cellular hairs arise from the epidermal cells.
3. Below the epidermis is a few layers sclerenchyma constituting the hypodermis.
4. Cortex is made up of parenchyma with large air cavities.
5. sphaerarophides are found in the cortical cells
6. Internal to the cortex is the endodermis.
7. Pericycle is found internal to the endodermis
8. Vascular bundles are many in number arranged in the form of a ring surrounding a central pith.

**Trapa (Trapaceae)**

1. Trapa is a free floating hydrophyte grows in open tanks and lakes.
2. Leaves are dimorphic floating and submerged. Floating leaves are large. They have a swolien petiole to help in buoyancy.
3. Submerged leaves arise in pairs at each node of the stem. They are highly dissected.
4. Stem is also submerged delicate and somewhat spongy.
5. Roots are unbranched and arise from the nodes.
6. Flowers are aerial but after pollination the pedicel bends down into water and the development of the fruit is under water.

**Fig 13.12 Trapa, Habit.**

9. Xylem is poorly developed. Phloem is found both external as well as internal to xylem.
10. Pith is wide and consists of parenchyma cells.

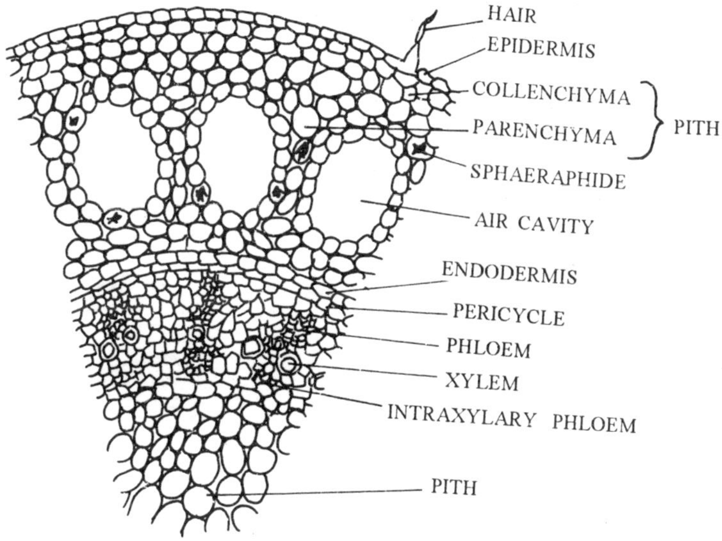


**Fig 13.13 Trapa, T.S. of Stem (A part enlarged)**

## Transverse Section of Leaf

1. Leaf is dorsiventrally flattened.
2. Upper epidermis consists of isodiametric cells. Cuticle is deposited above the epidermis.
3. Stomata are found only in the upper epidermis (leaves are epistomatic).
4. Lower epidermis is single layered and the cells are uninterrupted.
5. Mesophyll is distinguished into palisade and spongy parenchyma
6. Palisade cells may be in one or two layers below the upper epidermis. They have a number of chloroplasts.
7. Stomata in the upper epidermis lead to substomatal cavities present between the palisade cells.

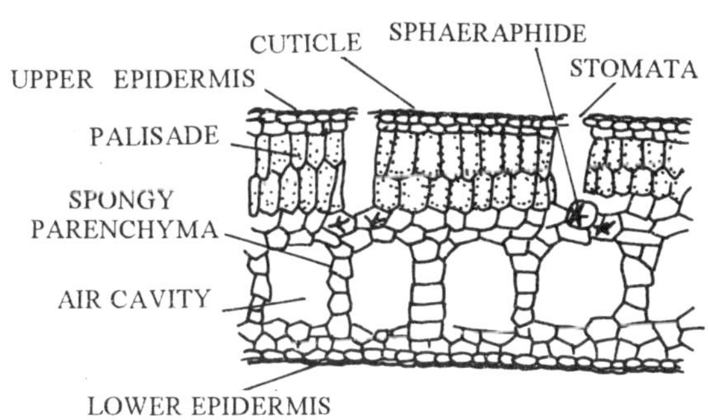


**Fig 13.14 Trapa, T.S. of Leaf.**

8. Cells present at the base of substomatal cavity have spaeroraphides.
9. Spongy parenchyma has large air spaces.
10. Vasculature is represented by a single vascular bundle in the midrib. Bundle sheath is parenchymatous.
11. Xylem is reduced to cavities. Phloem is well developed.

**Potomogeton (Hydracharitaceae)**

**Morphology**

1. The plant is a submerged rooted hydrophyte growing is shallow ponds and lakes.
2. Flowering branches are aerial.
3. Shoot is horizontal with vertical branches. Branches are thin delicate and long.
4. Roots are adventitious arising at the nodes
5. Leaves are semitransparent thin, thread like with a distichous phyllotaxy (arranged in two rows)
6. Mucilage covers all the underwater parts of the plant

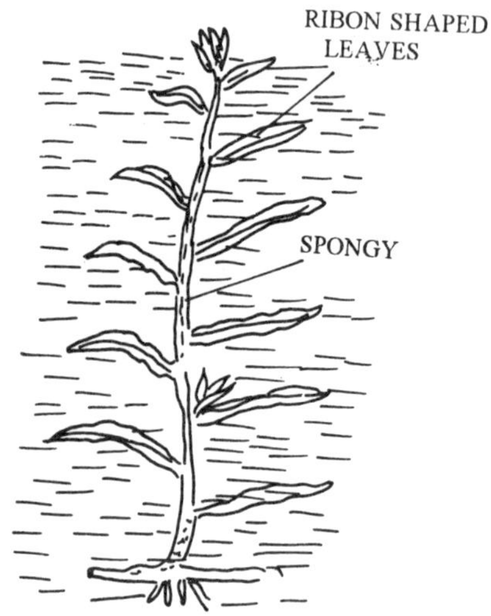

Fig 13.15 Potomogeton, Habit

**Transverse Section of Stem**

1. Transverse Section shows three regions - epidermis, cortex and stele.
2. Epidermis is single layered made up of barrel shaped cells. Cuticle is deposited above the epidermis.
3. Internal to the epidermis is a wide aerecnymatous cortex consisting of large air cavities separated by one celled partitions. Small patches of sclerenchyma are found in the cortex.
4. Endodermis is well developed while pericycle is indistinct
5. Xylem is reduced to a cavity while phloem is well developed.

**Transverse Section of Leaf**

1. Leaf is dorsiventrally flattened
2. Upper and lower epidermis are single layered arranged compactly.
3. Stomata are found only in the uppder epidermis
4. Each stoma leads to a sub stomatal cavity
5. Mesophyll is differentiated.
6. Palisade parenchyma is loosely arranged, so also spongy parenchyma.
7. Vascular bundle is surrounded by bundle sheath.
8. Xylem is highly reduced. Phloem is somewhat well developed.
9. At the margin leaf is just 3 celled in thickness.

**Eichhorinia (Potenderiaceae)**

**Morphology**

1. Plant is a free floating hydrophyte growing profusely in ponds and lakes.
2. There are a number of modifications in the root, stem and leaf to suit the aquatic habitat.
3. Stem is horizontal, cylindrical, spongy and is found just above the surface of water.
4. Stem breakes up into a number offsets which bring about vegetative propagation.
5. Roots arise in large clusters at the nodes, black in colour hairy and are called ballasting roots.

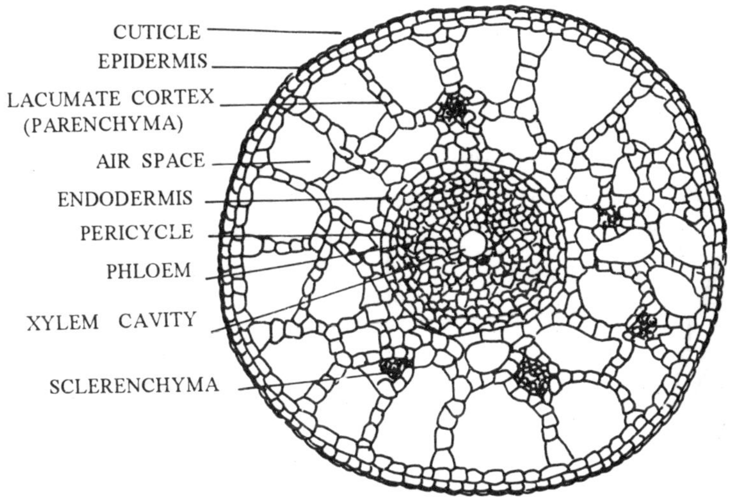


**Fig 13.16 Potomogeton, T.S. of Stem**

They help in balancing the plant on water.

6. Root caps are basent, instead large root packets are found.

7. Leaf has a spongy and bulbous petiole called phyllode. It helps in buoyancy.

8. Lamina is shiny due to the deposition of waxy material.

9. Pinkish flowers arise in clusters at the node

**Transverse Section of Petiole**

1. A T.S. shows epidermis and ground tissue in which the vascular bundles are scattered.

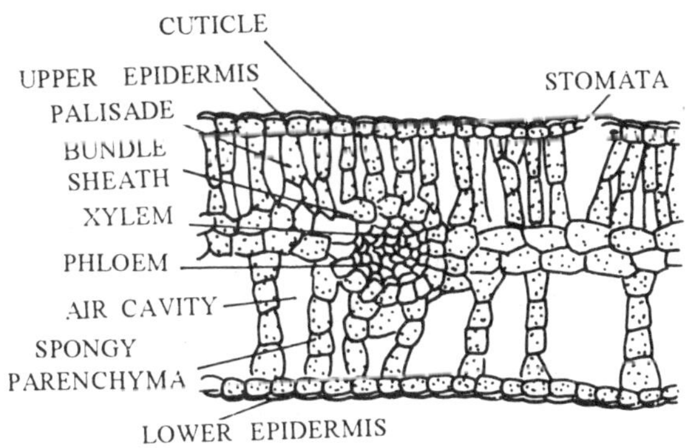


**Fig 13.17 Potomogeton, T.S. of Leaf**

2. Epidermis is made up of a single layer of thin walled cells. Cuticle is deposited over the epidermis.

3. Internal to the epidermis is the hypodermis made up of a few layers of compactly arranged cells. Cells are chlorophyllous.

4. Internal to the hypodermis is the ground tissue made up of aerenchyma with large air chambers separated by one cell thick partitions.

5. Dispersed in the ground tissue are found a number of vascular bundles. Xylem is poorly developed while phloem is well developed.

**Transverse Section leaf (lamina)**

1. Leaves are dorsiventally flattened
2. Both upper and lower epidermis are made up of a single layer of cells each.
3. Cuticle is deposited above the lower epidermis
4. Stomata are found in both the upper and lower epidermis (amphistomatic)
5. Mesophyll is differentiated.
6. Palisade parenchyma is one or two layered
7. Spongy parenchyma has large number of air cavities.
8. Vascular bundles have a prominent bundle sheath. Xylem however is poorly developed while phloem is well developed.

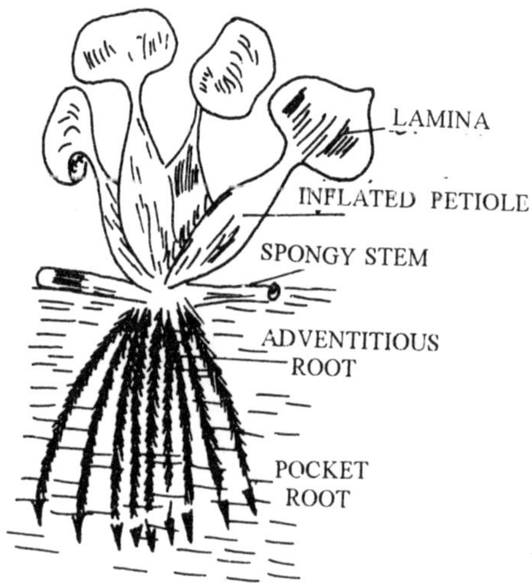


**Fig 13.18 Eichhornia, Habit**

**Transverse Section of Root**

1. A Transverse Section shows the three regions - epidermis, cortex and stele.
2. Epidermis is single layered made up of compactly arranged barrel shaped cells.
3. Internal to the epidermis is a few layered hypodermis followed by a broad aerenchymatous cortex.
4. Stele is surrounded by endodermis and pericycle
5. Vascular bundles are radial with exarch xylem. Xylem however is not well developed.
6. In the centre of the stele is present a parenchymatous pith.

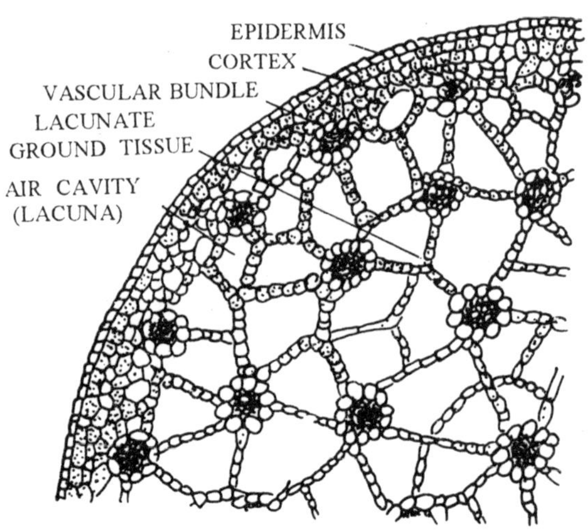


**Fig 13.19 Eichhornia, T.S. of Petiole (A part enlarged)**

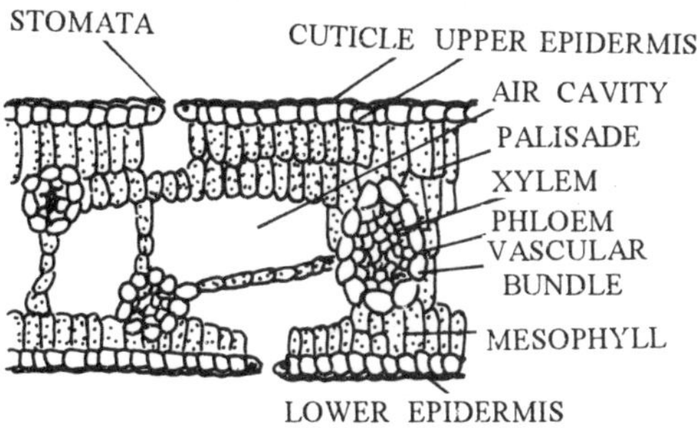


**Fig 13.20 Eichhornia, T.S. of Petiole (A part enlarged)**

**Vallisnaria (Hydrocharitaceal)**

1. Vallisnaria is a submerged rooted hydrophyte found growing in stationary or running water.
2. Roots are adventitious arising at the nodal regions of the horizontal underground stem.
3. Leaves are flat ribbon shaped and thin arising in a rosette fashion from the stem.
4. Plants are dioecious. In female plants the flowers are solitary and have a long coiled pedicle by the virtue of which the flower reaches the surface of water.
5. In the male plants flowers are in umbels which detach from the plant body and reach the surface of water.
6. Pollination is hydrophyllous (ephydrophyllous) and takes place on the surface of water.

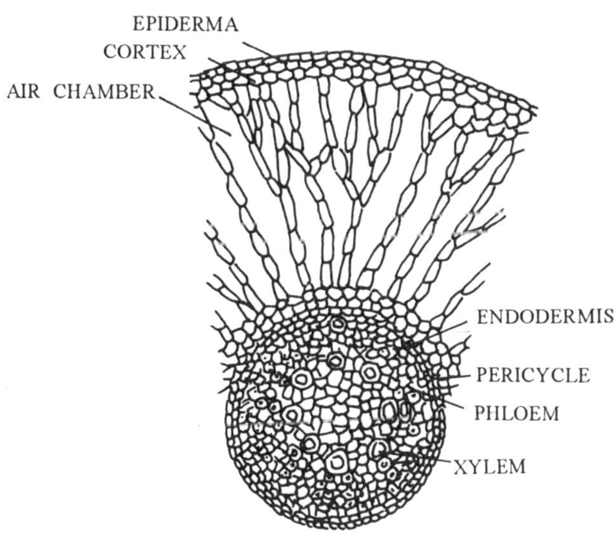


**Fig 13.21 Eichhornia, T.S. of root (A part enlarged)**

**Transverse Section of Leaf**

1. The leaf is isobilateral
2. Upper and lower epidermis are made up of a single layer of cells each. The cells have chloroplasts.
3. Cuticle and stomata are absent
4. Mesophyll is undifferentiated and consists of loosely arranged parenchyma cells with large number of air spaces.
5. Vascular bundles are reduced, Xylem is represented by a cavity while phloem is comparatively well developed.

**Jussieua (Onagraceel)**

**Morphology**

1. Plant is a free floating hydrophyte found in lake, ponds and ditches
2. Stem is horizontal growing just beneath the surface of water. It is distinguished into nodes and internodes. Stem is soft and spongy.

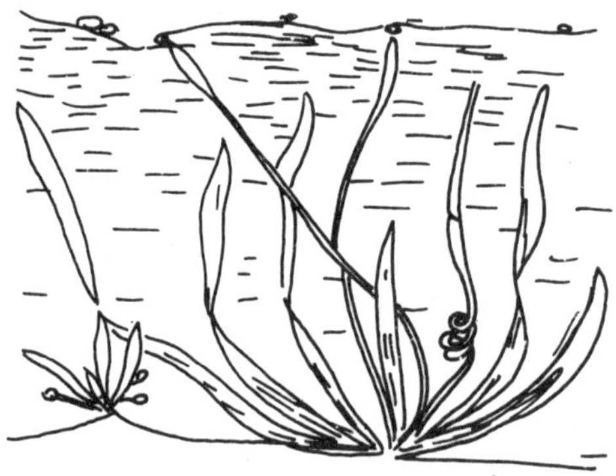

Fig 13.22 Vallisnaria, Habit

3. There are two types of roots - floating roots and ordinary roots. Ordinary roots are thin, fibrous and grow into water. Floating roots are white in colour spongy, soft short and stumpy arising in clusters of 3-4 at every node. They help in buoyancy.

4. In the anchorage roots, root hairs are absent, instead root packets are present.

**Transverse Section of Stem**

1. Transverse Section shows epidermis, cortex and stele.

2. Epidermis is made up of thin walled compactly arranged, radially elongated cells. Cuticle is absent

3. Next to the epidermis is the cortex divided into three zones. - outer cortex, middle cortex and inner cortex.

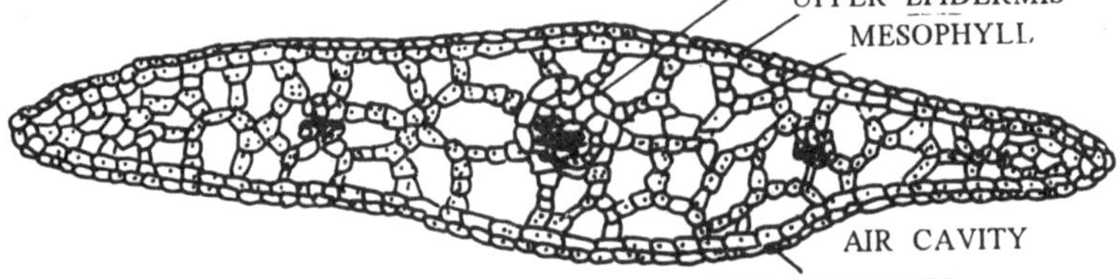


Fig 13.23 Vallisnaria, T.S. of Leaf

4. Outer cortex is composed of thin walled parenchyma cells with few intercellular spaces.

5. Middle cortex is aerenchymatous with large number of air spaces. The air spaces are separated by 3 - 4 celled wide partitions.

6. Inner cortex is made up of 2-3 layers of compactly arranged parenchyma cells.

7. Stele is surrounded by an indistinct endodermis.

8. Vascular bundles are many in number arranged in the form of a ring surrounding a central pith.

9. Vascular bundles are collateral and conjoint with poorly developed xylem but comparatively well developed phloem.

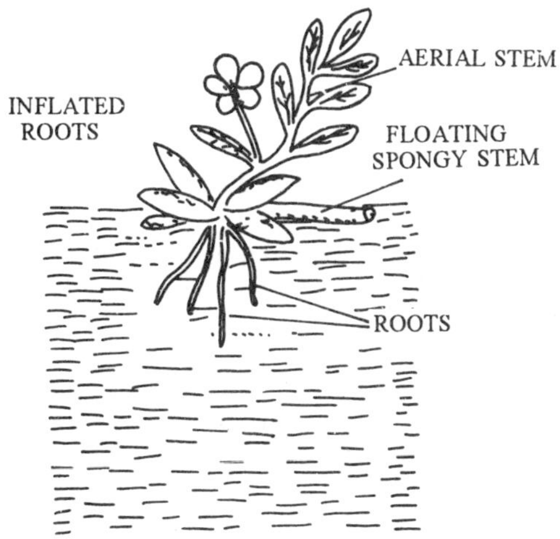


Fig 13.24 Jussieua, Habit

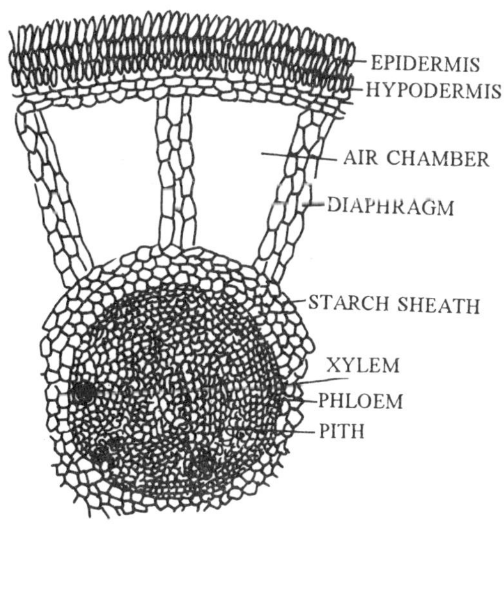


Fig 13.25 Jussiena, T.S. of Stem

**Ceratophyllum**

**Morphology**

1. Plants are submerged but rootless hydrophytes. They are feely suspended in water.
2. Stem is long, slender and delicate. Branches arise at the nodes.
3. Leaves are dichotomously dissected and arise in whorls at the nodal region. Margins of the leaves are serrate.
4. All the plant parts are covered with mucilage.
5. Mechanical breakage of the stem results in vegetative propagation.

**Transverse Section of Stem**

1. Transverse Section shows outer epidermis, middle cortex and the central stele.
2. Epidermis is made up of a single layer of thin walled compactly arranged cells. Cells are chlorophyllous.
3. Cortex is wide, chlorophyllous aerenchymatous with large number of air spaces.
4. Stele is surrounded by endodermis followed by a layer of pericycle.
5. Xylem is represented by an air cavity. Phloem is comparatively well developed.

**Transverse Section of Leaf**

1. The outline of cross section is almost circule.
2. Epidermis is single layered made up of compactly arranged thin walled chlorophyllous cells. Stomata are absent.
3. Mesophyll is undifferentiated composed of compactly arranged parenchyma cells. However 2-3 large air chammbers are present in the mesophyll.
4. Vasculare bundle is hihgly reduced and has poorly differentiated conducting tissue.

**Nelumbo (Lotus)**

1. Plants are rooted hydrophytes with floating leaves.
2. Leaves have a wide lamina with a long cylindrical soft air filled petiole.

3. Petiole and surface of lamina is coated with wax to prevent excessive absorption of water.
4. Some leaves may also rise above the surface of water.
5. Stomata are present on the upper epidermis.

**Pistia (Araceae)**

1. Free floating hydrophytes found on water surface in ponds, lakes etc.
2. Stem is horizontal, cylindrical and greenish in colour
3. Branches of stem called offsets help in vegetative propagation.
4. At each node on the stem a rosette of leaves arise in a spiral
5. Roots are adventitious arising in groups from the lower surface of the stem at the nodal regions.
6. Roots do not have root caps, instead they have root packets.

**Xerophytes**

Plants that grow in places where the availability of water is less are called xerophytes. Some xerophytes are also called desert plants. Xerophytes develop several morphological and anatomical adaptations to economise water.

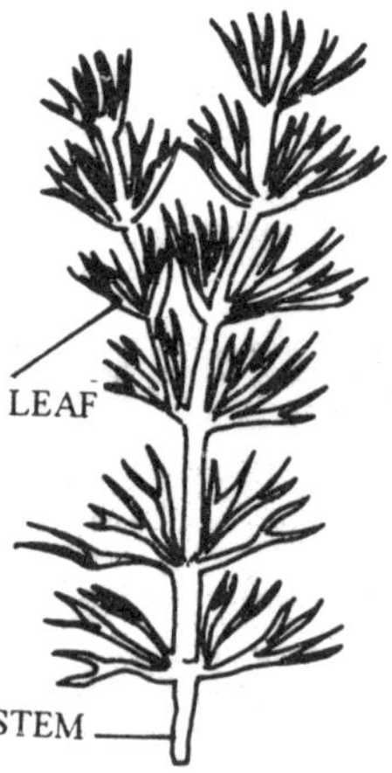


**Fig 13.26 Ceratophyllum, Habit**

XEROPHYTES

**Morphological adaptations**

1. Root system is well developed. It is both intensive and extensive for efficient water absorption.
2. Roots may be shallow but spread extensively when they are called surface feeders.
3. Roots are said to be deep feeders when they are not spreading but grow deeply into the soil.
4. Stems are stunted, thick and extensively branched.
5. In some cases stem gets modified into thick fleshy green structures called phylloclades or cladodes.

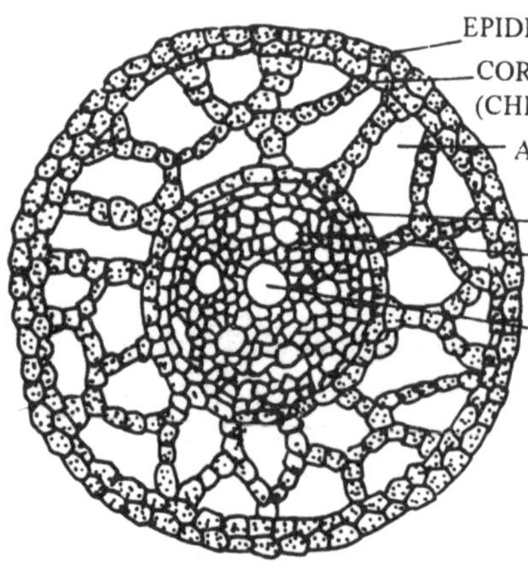


**Fig 13.27 Ceratophyllum T.S. of leaf**

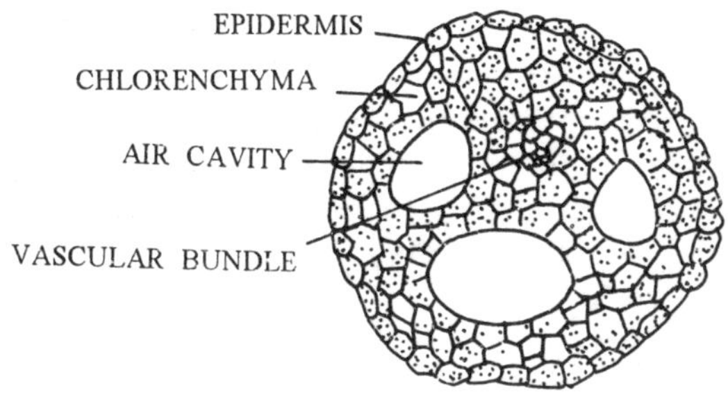


**Fig 13.28 Ceratophyllum T.S. of Leaf**

6. Stems may be underground in some cases or become hard and woody with thick barks.
7. Leaves are generally reduced, modified or totally eliminated.
8. Leaves may be modified into scales or spines which helps in checking transpiration as well as protects the plants from grazing animals.
9. In some plants leaves remain in a vertical posture exposing very little surface to direct sun light.
10. Some of the plants have decidous leaves which fall off during summer.
11. Leaves when present have a waxy or shiny surface to reflect right.
12. Leaves may be leathery or may have hairy outgrowths on the surface.
13. Under xeric conditions leaves may roll up and open only on the availability of water. (e.g. *Selaginella rupestris*).
14. Xerophytes produce large number of perenniating buds on the stem to help in vegetative propagation.

**Anatomical Adaptations**

1. Epidermis is well developed. Sometimes multiple epidermal layers are found.
2. Cuticle is very well developed.
3. Waxy or silica coating is often found on the aerial parts.
4. Hypodermal sclerencyma is often found beneath the epidermis in the leaves.
5. Stomata are reduced in number and usually leaves are hypostomatic

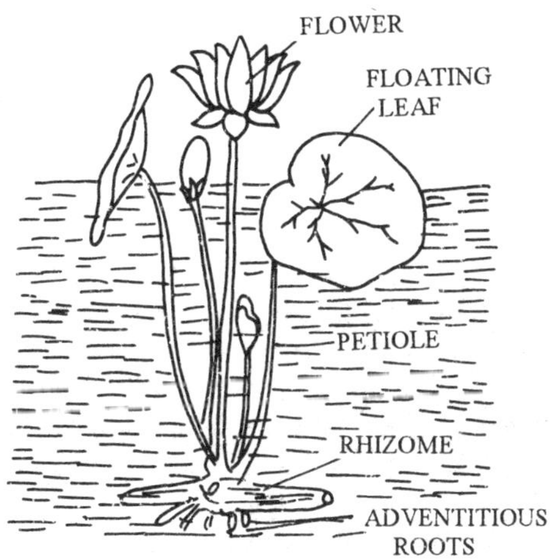


**Fig 13.29 Nelumbo, Habit**

6. In some plants stomata are scotoactive (i.e. open only during night).
7. Stomata are sunken, often covered by hairy outgrowths.
8. Many stomata are often found together in a sunken cavity (as in *Nerium* leaf).
9. Palisade parenchyma is very well developed; often it is many layered.
10. Air spaces are less in spongy parenchyma.
11. In succulent stems water is mixed with mucilage to prevent evaporation.
12. Mechanical tissues such as sclerenchyma, xylem etc are very well developed.
13. In some grasses motor cells are bulliform cells are present to help in the rolling up of the leaves.

We shall now study few examples of xerophytes

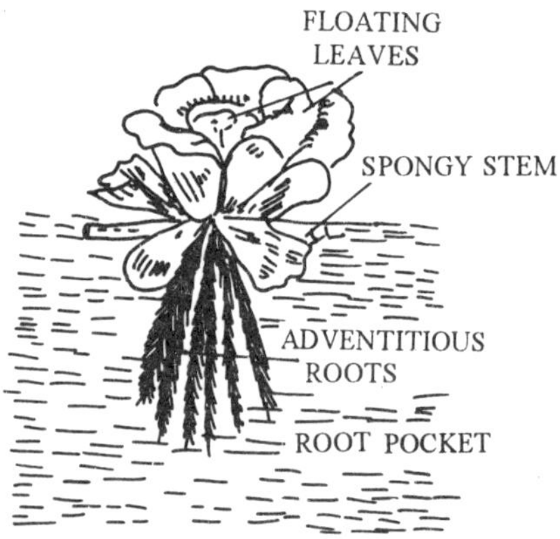


**Fig 13.30 Pista, Habit**

**Transverse Section of Stem**

1. Outline of the transverse section will be vaxy because of ridges and furrows.
2. Outermost layer is epidermis made up of a single layer of cells-cuticle is deposited above the epidermis.

**Casuarina equisetifolia**

**Morphology**

1. The plant is a tall tree.
2. Branches are green, jointed, look like leaves and many internodes long.
3. Leaves are scale like, highly reduced and present in a whorl at the nodal region.
4. Silica deposition is found on the young branches.

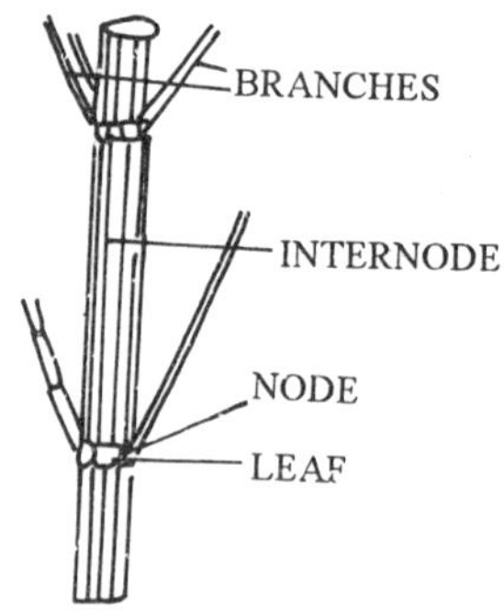


**Fig 13.31 Casuarina, a branch**

3. Stomata are sunken, restricted in distribution to the furrows.
4. Epidermal cells in the furrow have long multicellular hairs filling the cavity.
5. Next to the epidermis are found patches of sclerenchyma beneath the ridges.
6. Internal to the hypodermis the cortex is divided into ourter chlorenchyma and inner parenchyma
7. Chlorenchyma cells are arranged in many palisade layers constituting the chief photosynthetic tissue.
8. Embedded in the parenchymatous cortex are found a number of vascular bundles (one each below the ridge). These are called cortical vascular bundles and are collateral, conjoint, open with endarch xylem.
9. Next to the cortex is found the endodermis. Pericyle is represented by patches of sclerenchyma found between the endodermis and the stelar vascular bundles.
10. Vascular bundles are many arranged in the form of a ring surrounding a central pith. They are similar to the cortical vascular bundles in strcuture. Many phloem fibres are present in the vascular bundles.

**Calotropis gigantea**

**Morphology**

1. The plant body is a large herb or shrub.
2. All aerial parts are covered with white waxy material which acts as insulation against heat

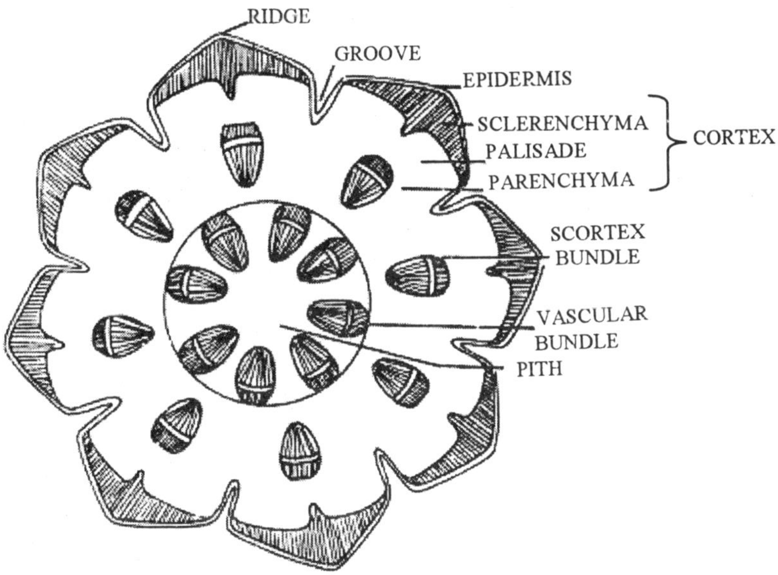


Fig 13.32 Casuarina, T.S. of Stem (Ground Plan)

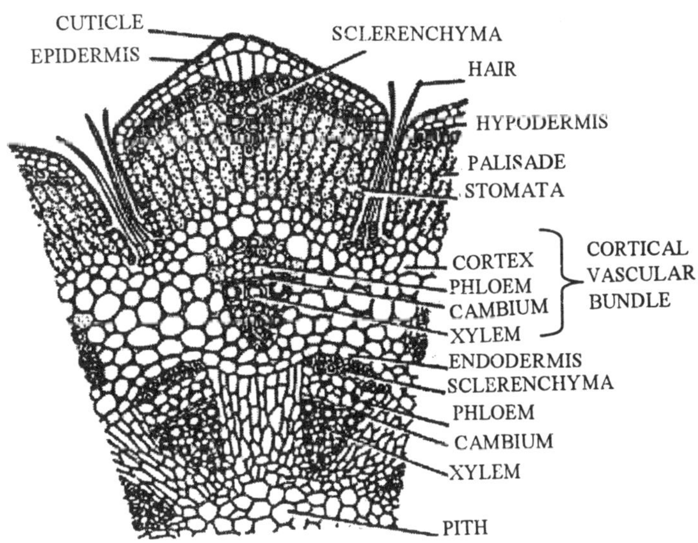


Fig 13.33 Casuarina, T.S. of Stem (A part enlarged)

3. Thick milky latex is present in the plant.
4. Leaves opposite and decussate. Inflorescence is a number.

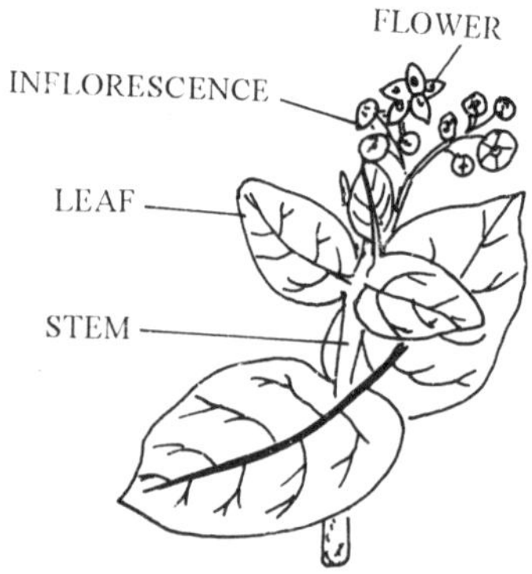


Fig 13.34 Calotropis, Habit

**Transverse Section of Stem**

1. Transverse Section shows epidermis, cortex and stele.
2. Epidermis is single layered. Thick cuticle and waxy deposition covers the epidermis.
3. Epidermal hairs present.
4. Cortex is divided into two zones - outer collenchymatous and inner parenchymatous.
5. Endodermis surrounds the stele. Pericycle is represented by patches of sclerenchyma.
6. Vascular bundles are many in number arranged in the form of a ring surrounding a central parenchymatous pith. They are bicollateral, conjoint, open with endarch xylem.
7. There are two groups of phloem in the vascular bundle (external and internal).
8. Numerous laticiferous vessels are seen in the pith.

**Transverse Section of Leaf.**

1. Leaf is dorsiventrally flattened.
2. Upper and lower epidermal layers are covered with waxy material.
3. Leaf is amphistomatic (stomata are found on both the surface).
4. Mesophyll is differentiated. Palisade cells are branched and loosely arranged.
5. Spongy parenchyma cells are also loosely arranged.
6. Vasculare bundles are covered with a prominent bundle sheath.
7. Laticiferous vessels are found in the mid rib regions.

**Euphorbia tirucalli**

**Morphology**

1. The plant body is a shrub.
2. Leaves are small in size and deciduous.
3. Stem performs the function of photosynthesis.
4. Milky latex is present in the plant body.

**Transverse Section of Stem**

1. A transverse section is nearly circular and shows the three regions - epidermis, cortex and stele.
2. Epidermis is single layered and is covered with thick cuticle.
3. Sunken stomata are found in the stem as the stem is the chief photosynthetic organ.
4. Cortex consists of a mixture of chlorenchyma and water storing parenchyma.
5. Latex cells are also present in the cortex.
6. Endodermis is not distinct, so also the pericycle.

7. Vascualr, bundles are arranged in a ring. They are collateral, conjoint and open with endarch xylem.
8. Pith is parenchymatous. Laticifierous cells present in pith also.

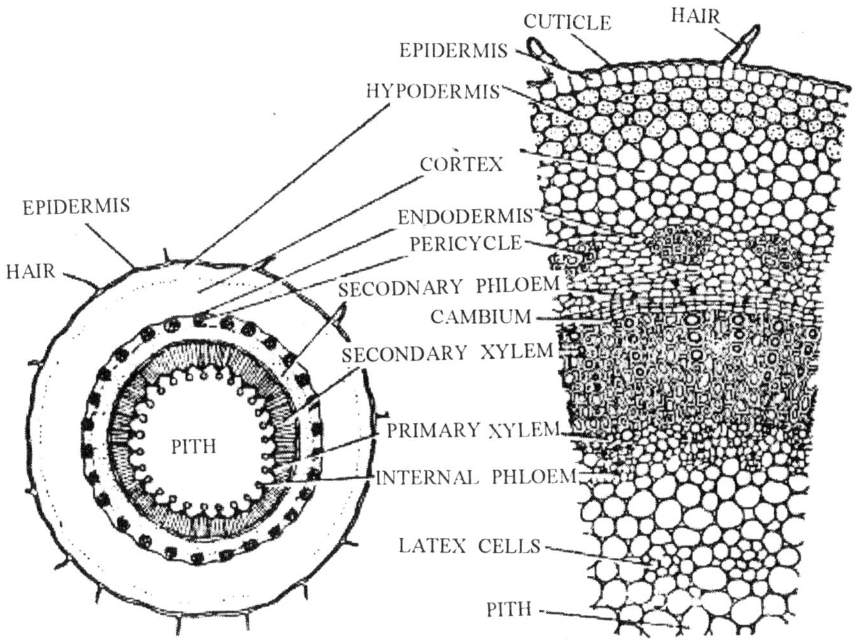


**Fig 13.35 Calotropis, T.S. of Stem**
**A. Ground Plan**
**B. A part enlarged**

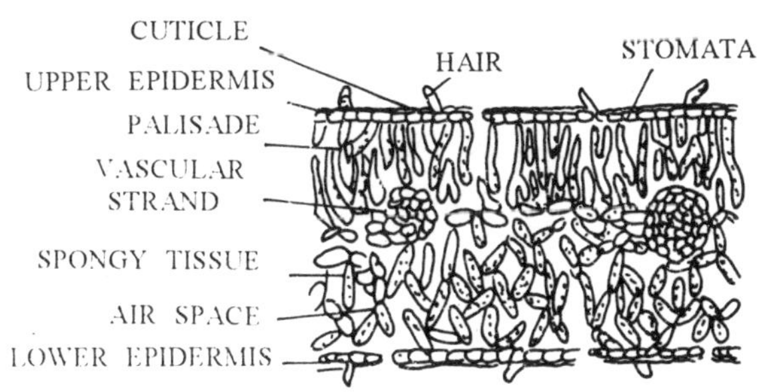


**Fig 13.36 Catotropis, T.S. of Leaf**

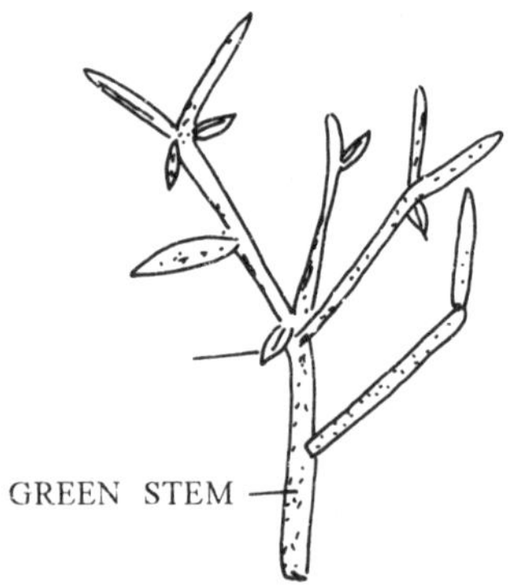


Fig 13.37 Euphorbia tirucalli, Habit

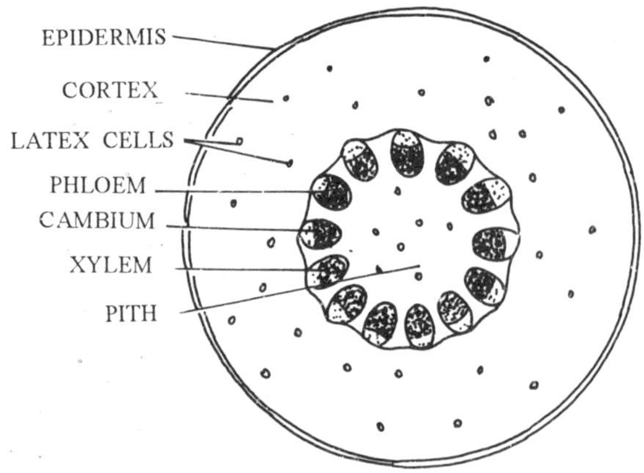


Fig 13.38 Euphorbia tirucalli, T.S. of Stem (Ground Plan)

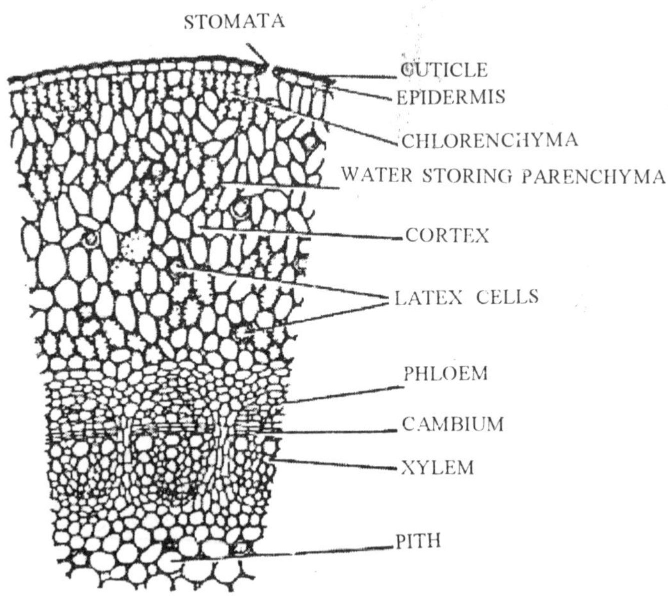


Fig 13.39 Euphorbia tirucalli, T.S. of Stem (A part enlarged)

### Asparagus

### Morphology

1. Plant body is a herb. Stem is cylindrical.
2. At the nodal region leaves and branches arise.
3. Leaves are reduced to scales.
4. In the axil of the scale leaves green flattened branches of one internodal length arise in clusters. These look like leaves. In reality they are modified branches know as cladodes.
5. As the leaves are reduced, cladodes take up the function of photosynthesis.
6. Roots are fasciculated, they are swollen into white tuberous structures. They store food material.

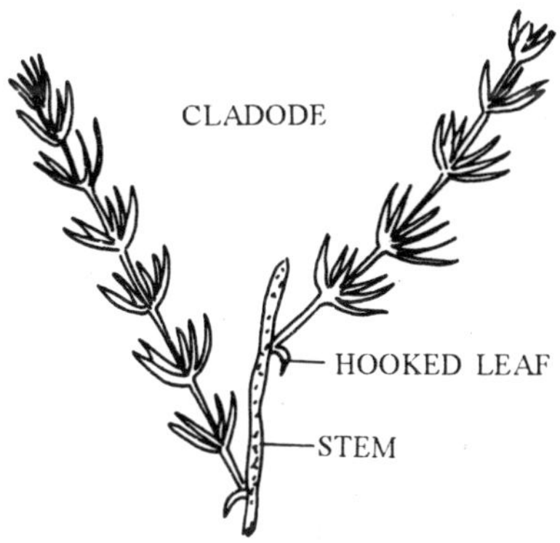


**Fig 13.40 Asparagus, Habit**

### Transverse Section of Stem

1. A transverse section has a wavy outline due to the presence of ridges.
2. Outermost layer is epidermis which is single layered. A cuticle is deposited outside the epidermis.
3. Next to the epidermis is found a few layers of chlorenchyma constituting the hypodermis
4. Internal to the hypodermis is the endodermis.
5. Pericycle is represented by a few layers of sclerenchyma.
6. Next to the pericycle is the ground tissue made upf of parenchyma cells with intercellular spaces.
7. Vascular bundles are embedded in this ground tissue. The bundles towards the central region are larger than those towards the periphery.
8. Vascular bundles lack a bundle sheath. They are conjoint, collateral and closed with endarch xylem. Xylem is 'V' shaped.

### Transverse Section of Cladode

1. The cladode is bifacial (dorsiventrally flattened).
2. Upper and lower epidermis each are single layered.
3. Towards the margins the two epidermal layers meet as the mesophyll becomes thin
4. Stomata are few in number
5. Mesophyll is differentiated, but the palisade is found all around the cladode internal to both upper and lower epidermis.
6. Internal to the palisade tissue is spongy parenchyma with many intercellular spaces.
7. There are two vascular bundles in the centre surrounded by thin walled closely packed parenchyma cells.

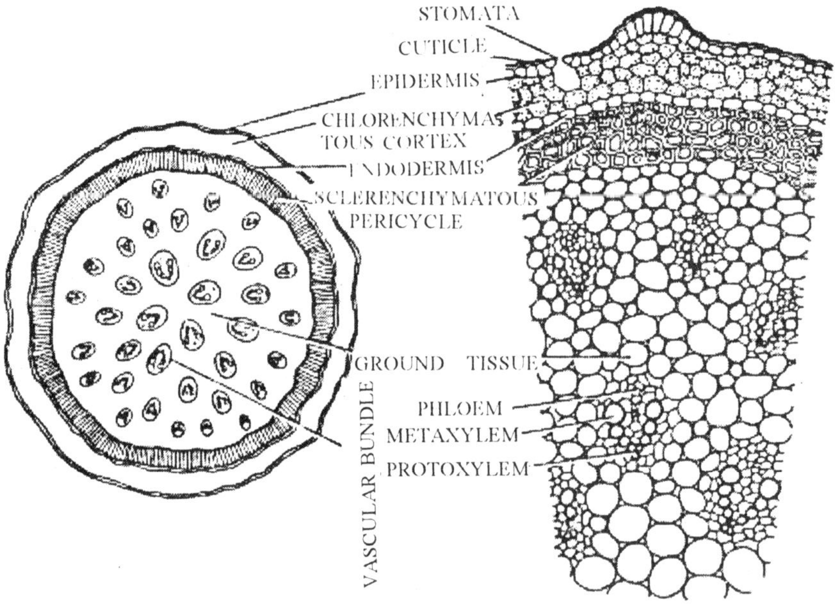


**Fig 13.41 Asparagus, T.S. of stem**
**A. Ground Plan**
**B. A part enlarged**

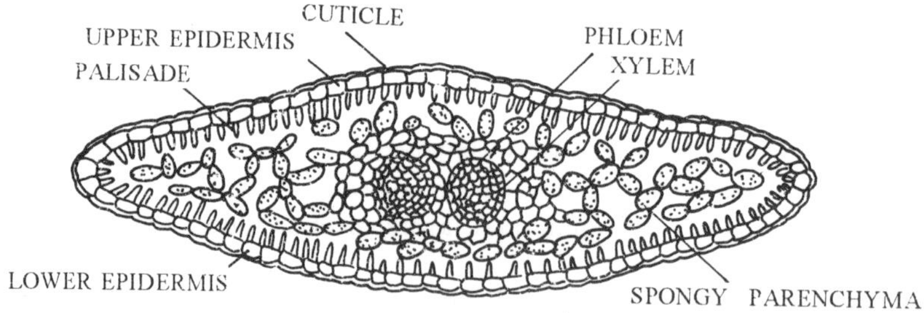


**Fig 13.42 Asparagus, T.S. of Cladode**

## CYNODON

### Morphology :

1. It is commonly called doob grass.
2. The plant usually grows in open fields, lawns etc.
3. The stem is a horizontal runner from which aerial shoots are prcduced.
4. Leaves are ribbon shaped, silica coated and have a long sheathing leaf base.

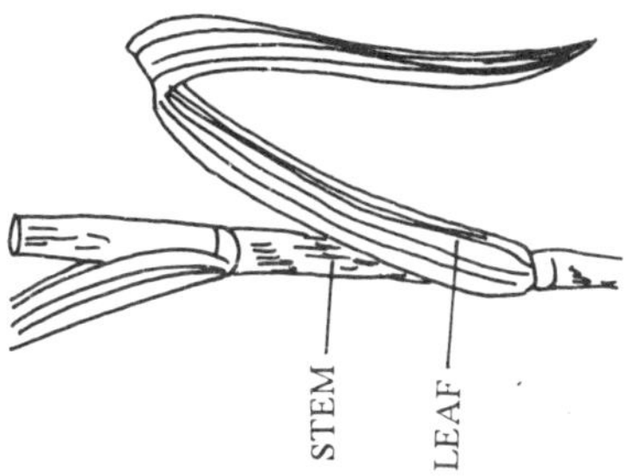


Fig 13.43 Cynodon, Habit

### Transverse Section of Stem

1. Outline of of transverse section is circular.
2. Epidermis is single layered with a deposition of cuticle above.
3. Next to the epidermis internally is found a few layers of chlorenchyma.
4. Next to the chlorenchyma is a band of sclrenchyma.
5. The entire region inner to sclrenchyma is filled by a mass of parenchyma cells constituting the ground tissue.
6. A number of vascular bundles are scattered in the ground tissue.
7. Each vascular bundle is surrounded by a sclrenchymatous bundle sheath.
8. Vascular bundles are conjoint, collateral, and closed with endarch xylem. Xylem is 'V' shaped. A cavity is formed by the dissolution of protoxylem elements. Phloem is situated above the xylem.

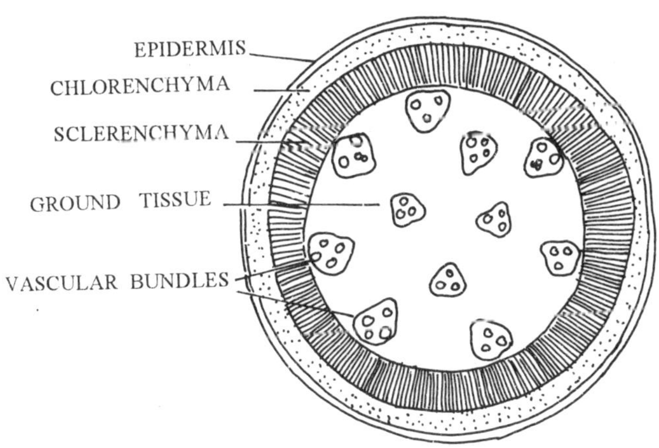


Fig 13.44 Cynodon, T.S. of Stem Ground plan

### Transverse Section of Leaf

1. Leaf is isobilateral.
2. Upper and lower epidermal cells are single layered and surrounded by cuticle.
3. Stomata are found in the upper epidermis.
4. There are certain special cells in the upper epidermis called motor cells or bulliform cells. These cells help the leaf to roll up during dry conditions
5. Mesophyll is undifferentiated Chlorenchyma cells with little or no intercellular speces constitute the mesophyll tissue.

6. A number of vascular bundles are arranged paralleyly (venataion is parallel). Xylem and phloem are well developed. Occasionally pads of sclerotic cells abutt the vascular bundles both above and below.

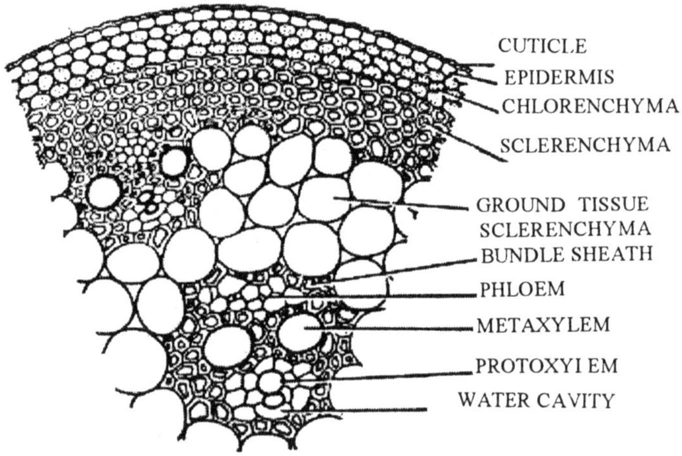


**Fig 13.45 Cynodon, T.S. of Stem**
**(A part enlarged to show the details of vascular bundle)**

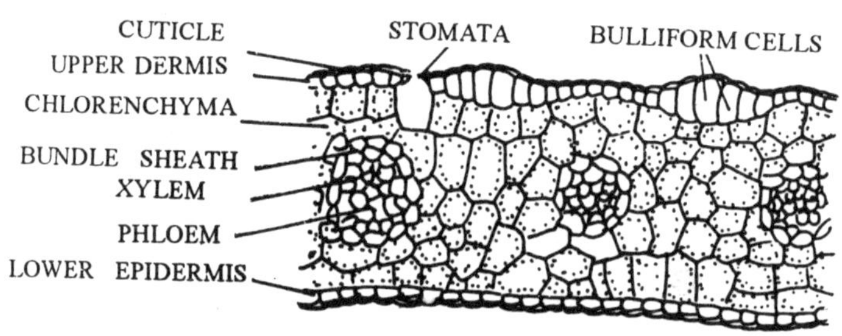


**Fig 13.46 Cynodon, T.S. of Leaf**

## MUEHLENBECKIA

### Morphology

1. Plant is a xerophytic shrub.
2. The entire stem is modified into a green flattened ribbion like structure called phylloclade.
3. Leaves are absent normally.
4. Occasionally small leaves arise from the nodal regions of the phylloclade
5. Stem takes up the function of photosynthesis
6. Flowers are borne in clusters at the nodal region.

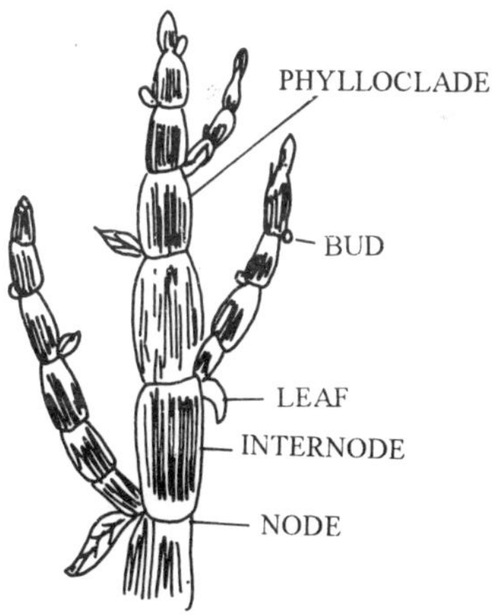


Fig 13.47 Machlenbeckia, Habit

### Transverse Section of Phylloclade

1. The phylloclade is dorsiventally flattened.
2. Upper and lower epidermis are made up of a single layers of cell each. Cuticle is found above the epidermis.
3. Internal to the epidermis all round there are patches of sclrenchyma.
4. In between the sclerotic patches chlorenchyma cells are found.
5. Endodermis is irregualr. Cells have starch grains. Pericycle is internal to the endodermis.
6. Vascular bundles are many in number; they are collateral open and endarch.
7. There is a parenchymatous pith in the centre.

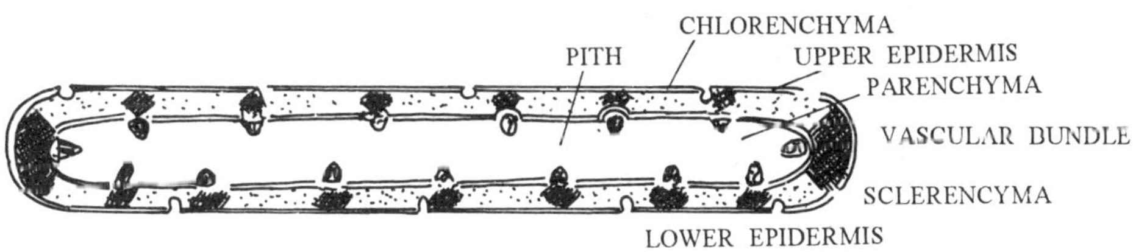


Fig 13.48 Muehlenbeckia, T.S. of Phylloclade (Ground Plan)

## RUSCUS

### Morphology

1. The plant is a tall herb or shrub.
2. Stem is cylindrical hard and green.
3. Branches are modified into the green flattened leaf like structures called plylloclades. Phyllocladeus are two internodes long. At the middle of the phylloclade (node) a small inflorescence arises proving the stem nature of the phylloclade.
4. Leaves are reduced to scales.

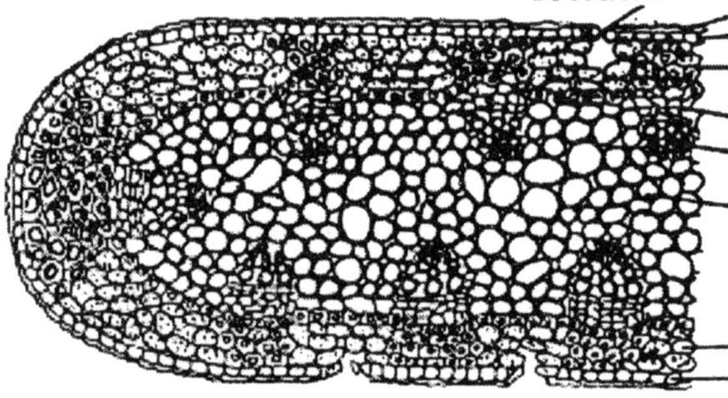


**Fig 13.49 Muehlenbeckia, T.S. of Phylloclade (A part enlarged)**

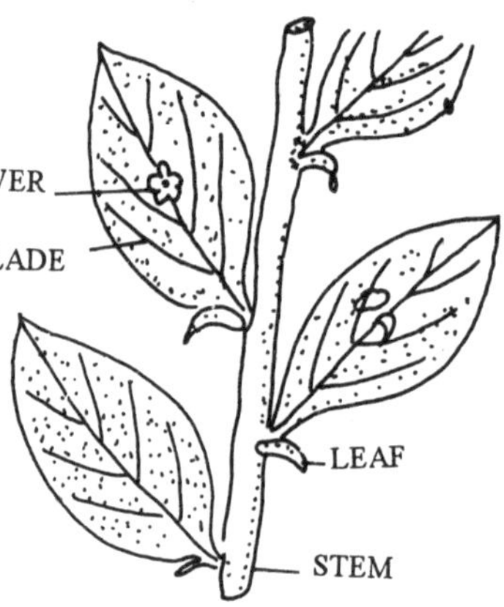


**Fig 13.50 Ruscus, Habit**

### Transverse Section of Phylloclade

1. Epidermis is the outermost layer. It is covered by a deposition of the cuticle.
2. Stomata are present both on the upper and lower epidermis.
3. Below the upper epidermis is found a few layers of chblorenchyma to help in photosynthesis.
4. The remainder of the phylloclade has parenchyma tissue.
5. The central region consists of the vascular bundles. The vascular bundles are amphivasal (phloem surrounded by xylem) and group of parenchyma cells are found in the centre constituting the pith region. All the vascular bundles are surrounded by a common bundle sheath.
6. Additional vascular bundles are found in the 'wing' region of the phylloclade. They are also amphivasal.

### Opuntia

1. The entire plant body (stem) of *Opuntia* is modified into a phylloclade.
2. Leaves are absent in the adult plant. Young plants have small caducous leaves.
3. The phylloclade is flat, green, jointed, branched and succulent. It performs the function of photosynthesis.
4. Internally the phylloclade has lots of mucilage
5. On the phylloclades spines are found which represent the modifications of leaves on the axillary branches.
6. Large, red coloured attractive flowers are borne at the nodes.

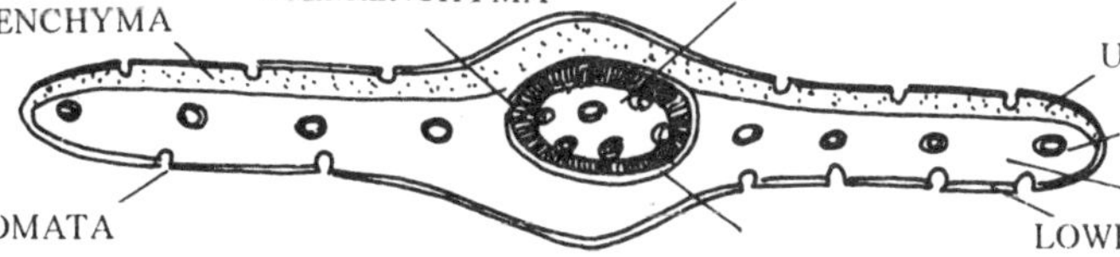


**Fig 13.51 Ruscus, T.S. of Phylloclade (Ground Plan)**

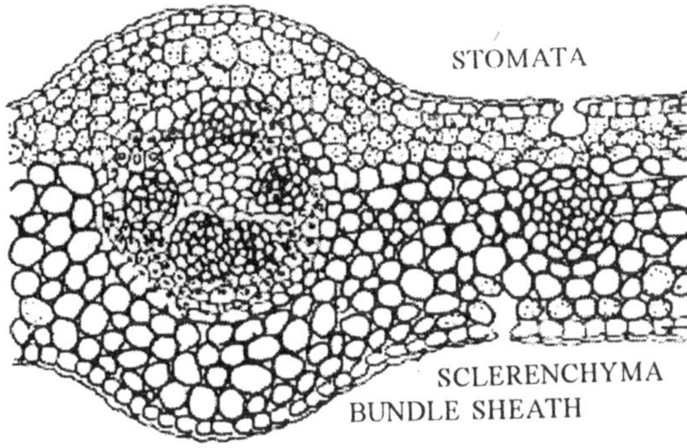


**Fig 13.52 Ruscus T.S. of Phyloclade (A part enlarged)**

## HALOPHYTES

Plants which grow and flourish in saline soils are called halophytes. In saline or marshy soil, salts of sodium chloride, magnesium sulphate and magnesium chloride are abundant.

### Rhizophora

#### Morphology

1. The plant body of *Rhizophora* is a small tree growing in the backwaters of sea.
2. Stem is anchored to the loose soil with the help of stilt roots.
3. In addition to the stilt roots certain special kind of roots grow apogeotropically and come above the water level. These are called pneumatophores or breathing roots. The pneumatophores have small pores called air pores which absorb air and provide the roots below as the soil has poor aeration due to water logging.
4. The seeds find it difficult to germinate in saline water due to osmotic imbalance. In order to overcome this the seeds germinate *in situ* i.e., they develop into small seedlings while still attached to the parent plant. This phenomenon is called Vivipary.
5. The seedlings get detached from the plant and fall straight into water and get dispersed due to water current.
6. Halophytes exhibit many xerophytic features such as, bushy growth, thick leathery leaves, hairy outgrowths, sunken stomata etc. This is due to physiological dryness.

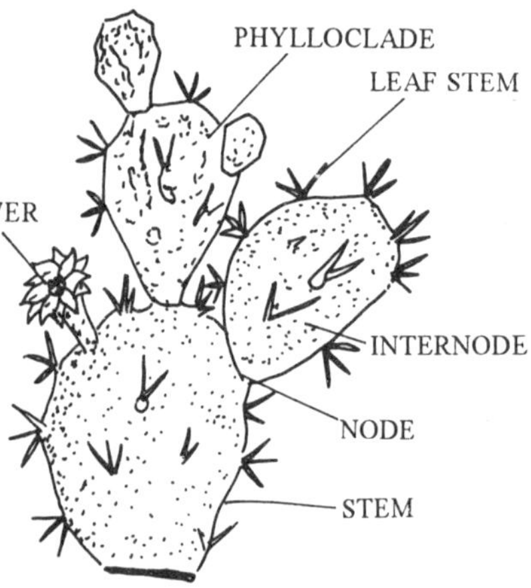


Fig 13.53 Opuntia, Habit

**PARASITES.**

Parasitic plants are achlorophyllous. They depend on green plants for their nutrition. Parasities are of two types - partial and total. Partial parasites depend on the host only for some of their requirements while total parasites are wholly dependent on the host plant. A few examples are discussed below.

**Viscum album**

1. This is a small herb a partial parasite growing on the branches of trees.
2. The stem is jointed. Leaves arise at the nodal region.
3. The stem produces haustoria which penetrate the host stem and absorb nutrition from the stem.
4. Seeds are sticky and when birds leave it on twigs (after eating the fruit), they (seeds) germinate there with the radicle coming out as a sucker
5. The sucker after penetrating the host tissue gives out branches which form an internal absorbing system.

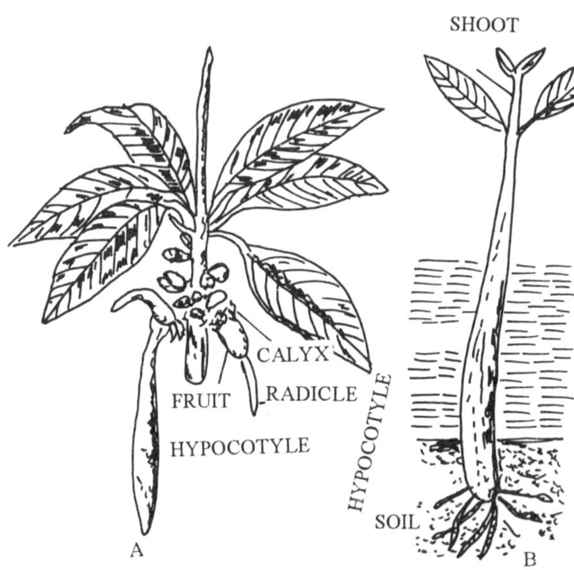


Fig 13.54 Rhizophora A. A branch showing Vivipary B. A seedling Establishing in the soil

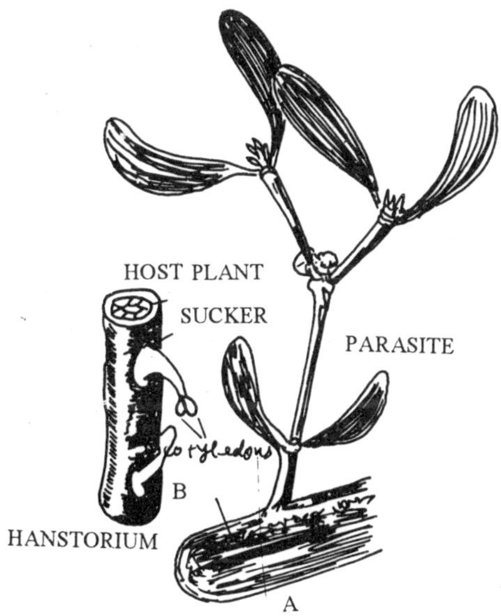


Fig 13.55 Viscum, A. Habit
B. A Stages of Germination on host stem.

**Cassytha filiformis**

1. The plant is a partial stem parasite.
2. The plant body consists of wixy stem that twines round the host stem.
3. Leaves and roots are absent.
4. Haustoria produced at several points from the stem penetrate the host stem and absorb nutrition.
5. The stem is greenish in colour and hence it can carry out some amount of photosynthesis.

Fig 13.56 Cassytha filformis
A partial Stem parasite (Habit)

**Santalum album**

1. This is a famouxs oil yielding tree.
2. Leaves are green in colour.
3. The plant is a partial root parasite.
4. Some of the lateral roots establish organic contact with the roots of other plants and absorb nutrition.

**Cuscuta reflexa**

1. The plant is a total stem parasite growing on host plants such as *Tecoma*.
2. The plant body consists of a yellow coloured stem that twines round the host stem.
3. Leaves and roots are absent.
4. Haustorial roots produced by the parasite penetrate the host stem reach the vasuclar bundles and absorb nutrition.

**Orobanche**

1. The plant is a total root parasite.
2. It has no leaves and grows on the roots of host plants.
3. The seed germinates only on coming into contact with the root of a host plant and forms an underground tuber.
4. The tuber gives rise to an aerial scape producing flowers which are birghtly coloured.

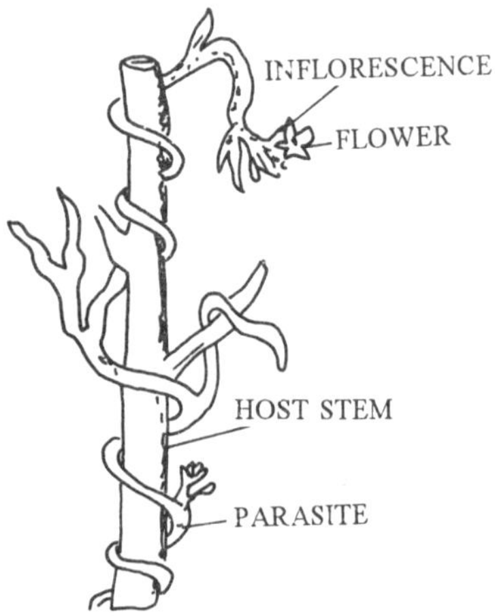


Fig 13.57 Cuscuta - A total stem Parasite, Habit

**Epiphytes**

Epiphytes grow on other plants only for shelter and sunlight. They are also called aerophytes. A number of plants belonging to families orchidaceae, Araceae, Cactaceae are epiphytic. *Vanda* one of the common orchid is discussed here.

**Vanda roxburghii**

1. The plant is a small herb growing on the branches of trees.
2. There are two types of roots - clinging roots and aerial roots (epiphytic roots).
3. Clinging roots anchor the epiphytes to the branch of the host plant.
4. Aerial roots hang down in air and absorb atmospheric moisture with the help of a tissue called velamen. This is the only source of water for the epiphyte as it does not absorb any water or nutrition from the host.
5. The epiphytic roots are green in colour and can carry out photosynthesis. Hence they are also called assimilatory roots.
6. Epiphytes develop many xeromorphic features as they get very little water. These are stunted growth, leathery leaves, thick cuticle, sunken stomata etc.

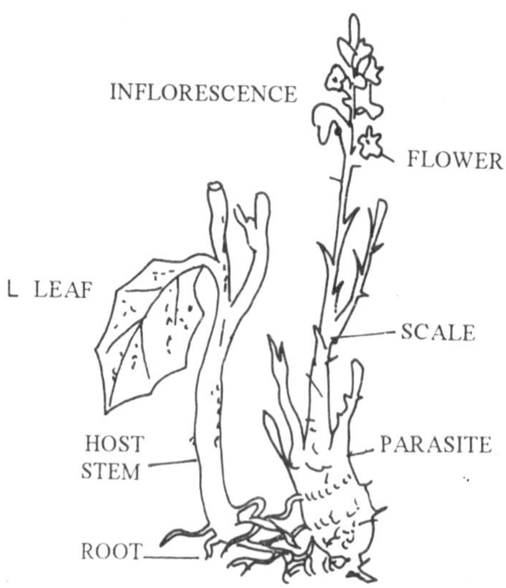


Fig 13.58 Orobanche, A total root parasite, Habit

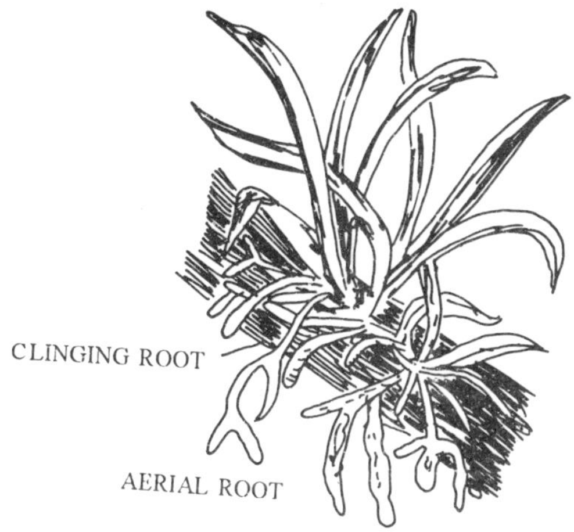


Fig 13.59 Vanda, an epiphytic Orchid

# PART FOUR
# PLANT PHYSIOLOGY

CHAPTER 14

# PLANT PHYSIOLOGY

Green plants in general and higher vascular plants in particular provide sustenance for all forms of life including human. Plants play a vital role in the survival of man on land. Human life as we know it perhaps would have been impossible but for plants. Plants provide us the three basic necessities of life namely food, clothing and shelter. In order to benefit from plant life, a thorough knowledge of their growth, development and function is necessary. For, it is this knowledge that can help us to make attempts towards improvement of plant life. The branch of botany that deals with various functional aspects of plants is called **Plant Physiology.**

Plant physiology may be said to be the study of vital phenomena in plants. It is concerned with the processes and functions, and also responses of plants to their surroundings. The result of plant's responses to the environment namely growth, development and reproduction are also studied in plant physiology.

The process studied in plant physiology are photosynthesis, respiration, translocation etc. Each process has a continuing sequence of events and plant physiology explains and describes each one of these events. Plant physiology also attempts to explain the functions of plants at various levels. Functions of cell organelles, cells, tissues and organs are explained.

To summarize, the following are the main tasks of plant physiology.

1. Describe and explain the process
2. Describe and explain the functions at various levels in the plant organization.
3. Explain the function of chemical constituents present in the body of plants.
4. Explain the effects of the environment of process and functions in the plant system.
5. Emphasize the interrelationship and interdependence of various processes and functions, so as to understand the life processes going on in theplant.

In order to realise the above objectives, plant physiologists derive help from physics, chemistry, mathematics, statistics etc. Indeed, the role of these physical sciences in greater in plant physiology than in other branch of biology. Modern developments in these branches of physical sciences have greatly helped in unravelling the secrets of many an important physiological process in plants.

Various tools and techniques are employed, both physical and biological, in the study of plant physiology. Microscope, chromatography, Autoradiography, Electrophoresis, Tracer techniques, Spectrophotometry and a whole range of biochemical techniques form a part physiology is a dynamic discipline which continuously interacts with other disciplines and constantly updates itself in tune with advancements in other disciplines of science. Developments in the field of biotechnology, genetic engineering etc., have also helped the study of various physiology processes *in vitro*.

# PLANT CELL

*1. Protoplasmic Streaming*

**Materials :** Leaves of *Elodea*, staminal hairs of *Tradescantia*, slides, cover slips, water, compound microscope.

**Procedure :** Collect fresh leaves or dissect out the staminal hairs. Place them on the slide and cover with the help of cover slip adding a drop of water. Observe under high power of the microscope, specially the chloroplasts in *Elodea* leaf.

**Observation :** The chloroplasts appear to move in a circulatory fashion around the vacuole.

In the case of *Elodea* there is a large central vacuole and the chlorplasts move in a clockwise or anticolock wise direction. This is known as Rotation. In the staminal hairs of *Tradescantia* there are many vacuoles and the contests move around each vacuole independently and it is known as Circulation.

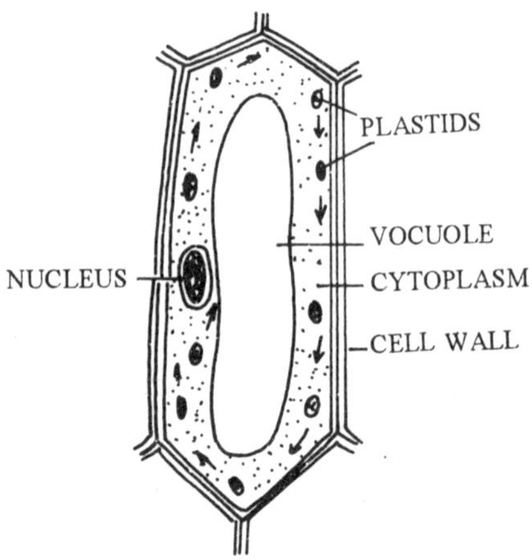


**Inference :** The movement of the chloroplasts seen under the microscope is actually due to the circulation or streaming of protoplasm. The movement probably occurs due to the sol-gel transformation of protoplasm. Microtubules of cytoplasm may also contribute to the movement.

*2. Importance of Living Membrane*

**Materials:** Beet root, cork borer, water, ice, spirit lamp, beakers, test tubes, thermometer, alcohol, water bath, test tube stand.

**Procedure :** Cut small cylinders of (approximately one c.m length) beet root with the help of a cork borer. Place the cylinders in a beaker and wash throughly under running water (to remove the pigments that ooze out) till the exudation of colour stops.

**Fig 14.1 Streaming of Protoplasm in Hydrilla (Leaf cell)**

Fill seven test tubes with about 20ml of water. to each test tube add 2 cylinders and keep the test tubes in water bath separately at different temperatures (25°, 30°, 40°, 50°, 60°, 75° and boiling temperature) for about 5 minutes. Also place two cylinders in ice cold water (20 ml) contained in a test tube for 5 minutes. Repeat the procedure with absolute alcohol and acetone.

**Observation :** Pigments from the cylinders of beet root start leaching out and the intesity of colour increases with increase in temperature with maximum being in boiling water. Increased pigment leaching is noticed in organic solvents like alcohol and acetone.

**Inference :** Under Normal temperature, membranes do not allow pigments to leach out, with increased temperature however permeability increases and in boiling water the membrane retentivity is lost resulting in maximum leaching. Alcohol and acetone treatments also result in membrane losing retentivity.

# SOLUTIONS, SUSPENSIONS AND COLLOIDS

3. *Preparation of 5% solution of sucrose (w/w)*

**Materials :** Sucrose, water, measuring cylinder, Balance etc.

**Procedure :** Weigh 5 gms of analytical grade sucrose and put it into a 100 ml volumetric flask. Add little water and stir to dissolve sucrose. Now make up the volume to 100 ml.

**Conclusion :** A true solution of sucrose is obtained by this procedure as the solute completely dissolves in water. This is a percentage solution weight by weight (w/w) i.e 5 gms sucrose in 95 gms of water

4. *Preparation of 5% solution of Sucrose (w1)*

**Materials :** Same as in experiment No. 3.

**Procedure :** Weight 5 gms of analytical grade sucrose and dissolve it in 100 ml of water. Stir until sucrose completely dissolved.

**Conclusion :** This is a percentage solution weight by volume (w/v) i.e., 5 gms of sucrose in 100 ml of water.

5. *Preparation of 80% alcohol (v/v)*

**Materials :** Absolute alcohol, distilled (water, measuring cylinder, beaker etc.)

**Procedure :** Take 80 ml of absolute alcohol and to it add 20 ml of distilled water. The volume will now be 100 ml.

**Conclusion :** This is a percentage solution of (liquid in liquid) volume by volume. The solution is 100 ml of 80% alcohol.

6. *Preparation of 1 Molar solution of Glucose*

**Materials :** Sucrose, distilled water, balance, measuring cylinder etc.

**Procedure :** Weigh 180.16 grams of glucose and dissolve it in some quantity of water. Then add water to make up the final volume to 1 litre

**Conclusion :** 1 molar solution of glucose will have 1 mole of glucose in one litre of water. What is a mole? Gram molecular weight of a substance dissolved in one litre of water will give one mole for every litre. What is gram molecular weight? Weight in grams of a substance equivalent to its atomic weight. In the case of glucose the atomic weight is $C_6 H_{12} O_6 = 12 x 6 + 12.6 + 16.6 = 180.16$x i.e., 180.16 gms of glucose in 1 gram molecular weight. Hence the above solution is 1 molar (1m). To prepare a 2 molar solution dissolve 360.32 gms of glucose in distilled water and make up the volume to one litre.

7. *Preparation of 1 molal solution of glucose*

**Materials :** Same as in experiment no - 6

**Procedure :** Dissolve a gram molecular weight of glucose (180.16 gms) in one litre of water.

**Conclusion :** This is slightly different from a molar solution in that the final volume of the solution will be more than one litre. Here one litre of a solution has slightly less than 1 mole of glucose per litre of the solution.

8. *Preparation of 1 normal solution of NaoH*

**Materials :** NaoH, Distilled water, balance, measuring cylinder etc.

**Procedure :** Weigh 40 gms of NaoH and dissolve it in some quantity of water. Make up the final volume to one litre.

**Conclusion :** In a normal solution a gram equivalent of a substance is dissolved in water to finally make one litre of the solution. What is gram equivalent weight? Gram molecular weight divided by the hydrogen equivalent. In NaoH the total molecular weight is 23 + 17 = 40 gms. Since there is only one hydrogen its value is 1. Hence 40 gms divided by 1 is 40. The gram equivalent weight of NaoH is 40. In this case gram molecular weight and gram equivalent weight are same as there is only one hydrogn.

### *9. Preparation of 1 normal $H_2So_4$*

**Materials :** Commercial, $H_2So_4$, Water, Measuring cylinder etc.

**Procedure :** Dissolve 49 gms or 28 ml $H_2So_4$ in one litre of water.

**Conclusion :** 49 gms is the gram equivalent weight of $H_2So_4$. Since it cannot be weighed it can be converted to volume by the formula V=M/D. Specific gravity of $H_2So_4$ is 1.84. Hence 49/1.8 = 26 ml. Since $H_2So_4$ is 8% pure about 28ml of H2SO4 is required to prepare 1 litre of 1N solution of $H_2So_4$.

### *10. Preparation of a Suspension*

**Materials :** Barium Chloride Solution, test tubes, dilute $H_2SO_4$, Test tube stand etc.

**Procedure :** Pour a little quantity of $BaCl_2$ solution in a test tube. Add a few ml of sulphuric acid. Shake the mixture thoroughly.

**Observation :** A white precipitate of barium sulphate settles down.

**Conclusion:** In a suspension, particles remain dispersed in the medium. They do not dissolve. Suspensions are generally unstable and gradually the particles settle down and the pure solvent can be separated.

### *11. Preparation of an Emulsion*

**Materials :** Groundnut oil or mustard oil, water, test tube etc.

**Procedure :** Take little amount of oil (10ml) in a test tube. Add 20 ml of water and shake vigorously.

**Observation :** Oil and water are imiscible. Water is polar while oil is non polar, further the density of oil being less than that of water, it (oil) floats on the surface of water. . Thus two layers (oil and water) are formed in this emulsion.

### *12. Demonstration of Brownian Movement*

**Materials :** Microscope, latex of *Calotropis*, or *Nerium* India ink, slide, cover slip etc.

**Procedure :** Place a drop of the latex on the slide and cover it with cover slip. Observe it under the high power objective of the microscope

Dilute India ink abou t10 times and follow the same procedure as for latex.

**Observation :** Small particles in the liquid get illuminated and move to and fro in the microscopic field.

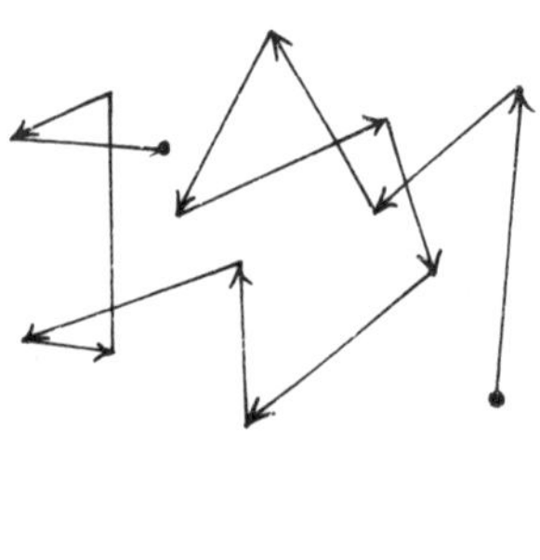

Fig 14.2 Demonstration of Brownian Movement

**Conclusion :** The particles are in constant motion due to their inherent kinetic energy in a colloidal system. Uneven bombardment of the particles by the dispersion medium results in this movement. This movement was first discovered by Brown, hence it is called Brownian movement.

### *13. Demonstration of Tyndall effect*

**Materials :** 1% glucose solution, fruit juice, beaker, pencil, torch, water.

**Procedure :** Take two beakers and pour glucose solution in one and fruit juice in the other. Keep the beakers in darkness. Direct the torch towards each beaker and switch on the light so that the beam of light passes through the solution or fruit juice. Observe the behaviour of the light beam in both the cases.

**Observation :** The beam of light will be normal in the case of beaker having glucose solution., while in the case of fruit juice the light beam gets scattered.

**Conclusion :** Fruit juice being a colloideal solution the light beam gets scattered as the particles are in constant motion. This phenomenon of scattering of light while passing through a colloidal solution is known as Tyndall effect.

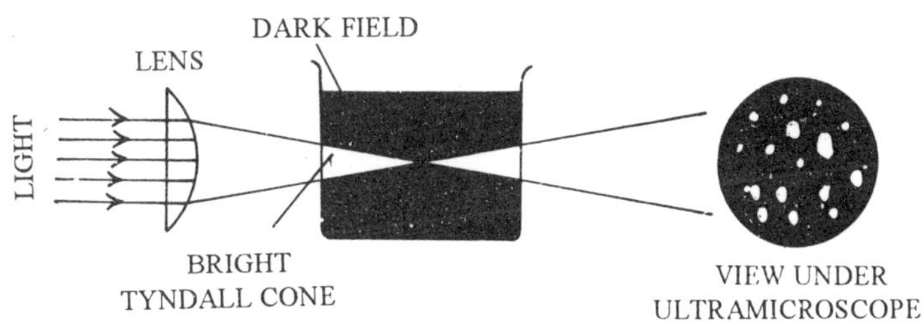


**Fig 14.3 Demonstration of Tyndall effect**

### *14. Separation of Starch and Salt by dialysis*

**Materials :** Collodion bag, starch, Nacl, water, test tube, spirit lamp, silver nitrate, iodine, thread, stand, glass rod, beakers etc.

**Procedure :** Prepare 2 % solution of Nacl and a colloidal solution of starch (Add 1 gm of starch to 50 ml of boiling water with constant strirring). Mix the two solutions in equal volumes. Fill the mixture in a dialysis bag (collodion bag) and immerse it as shown in the figure in a beaker containing distilled water. After about 1 or 1 1/2 hours, test the water of the beaker with silver nitrate and Iodine Solutions separately.

**Observation :** On treated with silver nitrate, the water turns milky showing the presence of salt. It does not turn blue idicating starch has not diffused out of the dialysis bag.

**Conclusion :** The collodion membrane is differentially permeable, wnile it allows the salts to go out, it does not allow starch to go out.

**Movement of Substances**

### *15. Demonstration of Diffusion :*

**Materials :** Beaker, water, $CuSo_4$, crystals.

**Procedure :** Place few crystals of copper sulphate in a beaker containing water and observe.

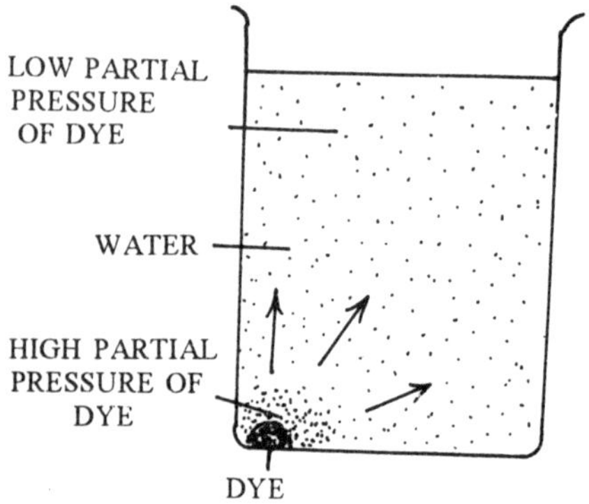


**Fig 14.4 Demonstration of Diffusion**

**Observation :** The crystals dissolve slowly in water. The particles may be seen moving away rapidly from the crystal to other parts in water. After the crystals completely dissolve this unidirectional movement stops and there will be to and fro movement.

**Conclusion :** The above experiment demonstrates the diffusion of solid particles. Diffusion is defined as the particles of solid, liquied organs distributing themselves evenly throughout the available place.

***16. Demonstration of Imbibition***

**Materials :** Dry seeds, water, pertridish.

**Procedure :** Soak the seeds in water overnight. Observe the ssize of the seeds. Check the termperature of water.

**Observation :** The seeds become swollen. After some time the seed coat breaks open. There will be slight rise in termperature of water.

**Conclusions :** Seeds swell due to imbition of water. The cotyledons swell at a much faster rate than seed coat, the seed coat breaks open exposing the cotyledons. This helps in germination. Imbibition involves both *absorption* as well as *adsorption*. Due to adsorption water molecules are held over the surface of the seed. As they cannot move they lose their energy in the form of heat into water; hence the rise in temperature of the water.

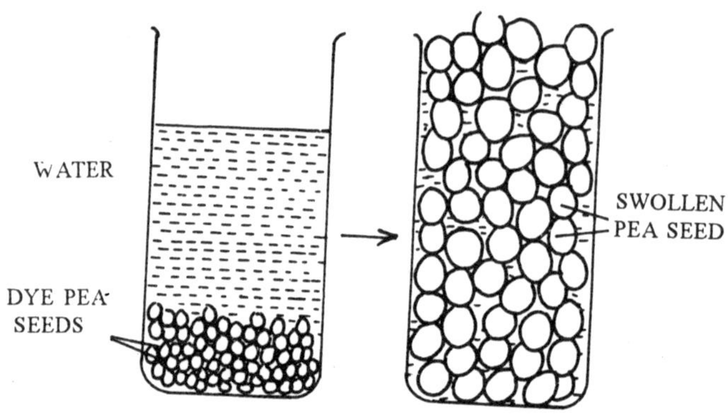


**Fig 14.5 Demonstration of Imbition**

***16. Demonstration of Imbition Pressure***

**Materials :** Imbibitional pressure instrument (disatometer), dry bean seeds, water and weights.

**Procedure :** Keep the apparatus as shown in the figure. The apparatus consists of a cylindrical jar fitted with a metallic piston. The metallic piston has a flat pan on its top to which weights can be attached. To the upperr end of the piston is atached a pointer (indicator) which glides above and below on a graduated arc scale fitted to a stand.

Keep some dry bean seeds in the apparatus and add water into it. Press the piston over the seeds by attaching weights on the pan. Note down the initial reading on the scale. Leave the experiment for about two hours. Note the final reading on the scale.

**Observation :** As the seeds soak in water they become swollen and push the piston above. As a result the indicator moves down on the scale certain degrees.

**Conclusion :** Due to imbibition the seeds swell and as they are confined in a closed container they develop a pressure and as a result the piston moves up. If we put enough weight on the pan so that the piston does not move up, that weight or force equals the imbibitional pressure developed by the seeds.

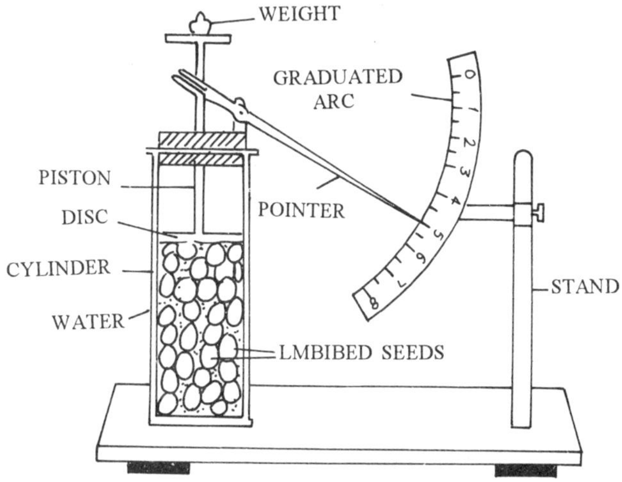


**Fig 14.6 Apparatus for measurement of Imbibition Pressure**

### *17. Physical Demonstration of Osmosis (Thistle Funnel Experiment)*

**Materials :** Egg membrane, animal bladder or parchment paper, 5% sugar solution, thistle funnel, beaker, water, stand etc.

**Procedure :** Take a beaker half filled with water. Take a thistle funnel and tie its mouth with egg membrane parchment paper or animal bladder. Fix the thistle funnel to a stand inversley so that the funnel is immersed in water. Fill the funnel with 5% sugar solution. Note the level by making a mark on the stem of the thistle funnel. Allow the experiment to stand for about 30 minutes. Then note the final level of sugar solution. Taste the water in beaker

**Observation :** The level of sugar solution in the thistle funnel has gone up. Water in the beaker does not taste sweet.

**Conclusion :** The egg membrane is selectively permeable only to water and not to sugar. Hence, as the solvent concentration is higher in beaker than in the thistle funnel. Water migrates into the thistle funnel and the level of sugar solution rises up. This phenomenon is known as osmosis which may be defined as Migration of solvent molecules from the region of *their* higher concentration to the region of *their* lower concentration across a selectively permeable membrane.

### *18. Physiological Demonstration of Osmosis (Potato Osmoscope)*

**Materials :** Potato tubers, 5% sugar solution, water, petridish, pins.

**Procedure :** Select a large size raw potato and carefully peel off the skin. Cut into two pieces horizontally.

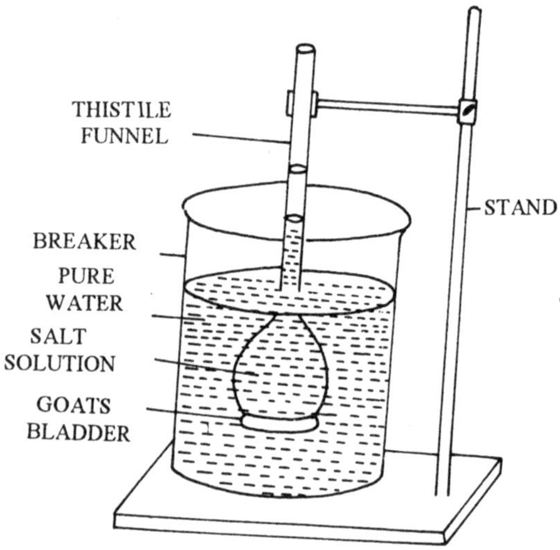


**Fig 14.7 Physical Demonstration of Osmosis (Thistle funnel experiment)**

Scoop out potato tissues from the centre and fashion a small cup as shown in the figure. Pour a little of sugar solution into the potato cup. Stick a pin inside the cup to note down the level of sugar solution. Keep the potato cup (it is now called a potato osmoscope) in the petridish containing water. Leave the apparatus aside for about 30 minutes. At the end of this period note the final level of sugar solution in the potato osmoscope and stick a pin to mark the level.

**Observation :** It is observed that the level of the sugar solution has gone up inside the potato cup due to entry of water from petridish.

**Conclusion :** Potato tissues act as a selectively permeable barrier and allow only water but not solute particles to pass through. As a result water which is in higher concentration in the petridish migrates to the potatocup. This migration is due to osmosis. (For definition of osmosis see the previous experiment).

This experiment is known as physiological demonstration of osmosis as it is seen in living tissues, while the thistle funnel experiment in physical demonstration because it is seen in membrane removed out of the living tissues.

### *19. Prepartions of a Semi permeable Membrane (Collodion bag)*

**Materials :** Collodion solution (Cellulose nitrate dissolved in alcohol and ether), water, test tube glass rod, toluol etc.

**Procedure :** Pour a little collodion solution into a test tube. Keep the test tube in a slanting position and revolve it slowly so that the solution forms a coating on the inner walls of the test tube. Decant the extra solution. Allow it to settle. Then give a second coating of the solution in the same manner. After some time the solution dries up into a membrane which can be carefully peeled off from the interior of the test tube.

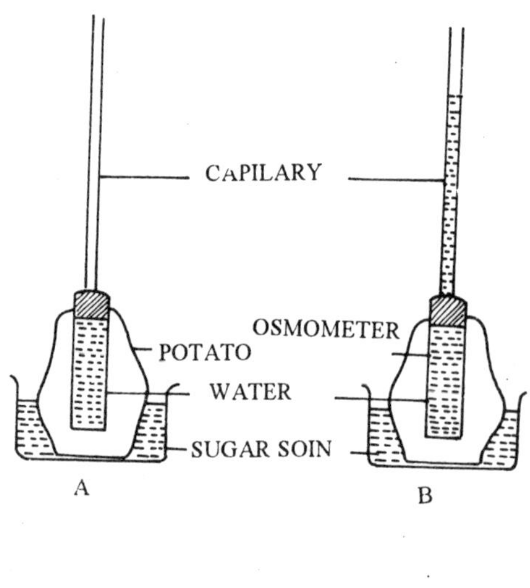


**Fig 14.8 Demonstration (Physiological) of Osmosis using Potato Osmoscope**

Collodion sac prepared as above can be preserved in water to which a few drops of toluol have been added. This membrane has the property of semipermeability

*20. Demonstration of plasmolysis*

**Materials :** Onions or Tradescantia leaf, safety blade, sugar solutions of different concerntrations, cover slips, slides, water, microscope etc.

**Procedure :** Onion peel can be obtained by tearing the onion leaves obliques when a thin transparent layer comes out. Semilarly leaf epidermal peels also can be obtained. Set up a series of slides and place a drop or two of water, 0.25 m, 0.5 m, 0.75 m and 1.0 m sugar solution on the slides as shown below.

| Slide : | 1 | 2 | 3 | 4 | 5 |
|---|---|---|---|---|---|
| | Water | 0.25m | 0.5m | 0.75m | 1.0m |

(Sugar Solution)

on each slide place a piece of the onion peel and put the cover slip. Observe the shape of the cell, cytoplasm etc and draw a sketch. Leave the preparation aside for 15 minutes. Observe again.

**Observation :** In slide No. 1 (water) the cells retain their original size and shape and in slide 2 (0.25m) there is slight shrinking of the cytoplasm. A small space will be left between the cell wall and plasma membrane. In slide No. 3 (0.5m) the cytoplasmic shrinkage is considerable and the plasma membrane becomes separate from the cell wall. In slides 4 and 5, cytoplasm completely loses contact with the cell wall and shrinks to its minimum size. The cytoplasm will be either concave or convex in shape.

**Conclusion :** In ordinary water, the cell retains its shape and size or slight increate might be there due to entry of water. Water is thus hypotonic in comparison with cell sap. In slides 2, 3, 4 and 5 the solution will be increasingly hypertonic as the shrinkage in cytoplasm becomes more due to exit of water. This exit of water from a living cell when placed in a hypertonic solution is known as plasmolysis.

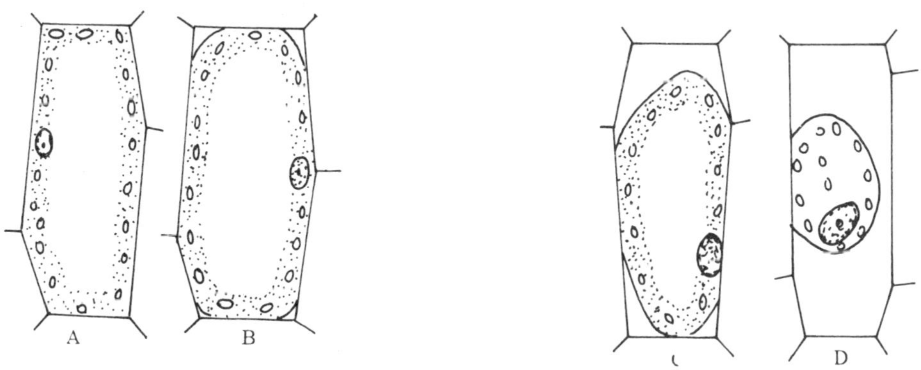


**Fig 14.9 Demonstration of Plasmolysis**

*21. Determination of Osmotic Concentration of Cell sap by Plasmolytic Method*

**Materials :** 1m glucose solution, onion peel, cavity slides, microscope etc.

**Procedure :** Prepare the following series of dilutions of glucose solutions from 1m glucose solutions.

0.1m, 0.2m, 0.3m, 0.4m, 0.5m, 0.6, 0.7m, 0.8m, 0.9 and 1 m

Take a few drops of the above concentration and place a piece of the onion peel. Observe the shape and size of cells. Leave them aside for 30 minutes and observe.

**Observation :** Note down as to the increase, decrease or no change of the cell size in different concerntrations of the solution.

**Conclusion :** The concentration of the solution where there in neither increase or decrease of the onion peel cells equals the concentration of the cell sap. This is the isotonic concentration. Wherever there is increase, it is hypotonic and wherever there is decrease it is hypertonic. For instance if 0.2m is the isotonic concentration osmotic pressure concentration) of cell sap can be calculated as follows.

OP = 22.4 atm x Molarity of external solution x Absolute temperature + Room Temperature (Approximately $24^{0}$C / 273

OP = 22.4 x 0.2 x 273 + 24 / 273

OP = 4.56

*22. Measurement of Diffusion Pressure deficit (Water potential)*

**Materials :** Potato tubers, cork borer, measuring cylinder 1m glucose solution, test tubes, balance, filter paper, scale, test tube stand.

**Procedure :** Prepare a stock solution of 1m glucose in distilled water (dissolve 180.16 gms in water to make the volume to 1000 ml) and prepare serial dilutions from it as follows.

0.1m, 0.2m, 0.25m, 0.3m, 0.35m, 0.4m and 0.5m. Take 15 ml of each of these dilutions in separate test tubes and keep them in a stand. Take 15 ml of distilled water in another test tube.

Peel the skin of potato and with the help of a cork borer scoop out cylindrical pieces of potato. Cut them into 24 pieces of equal length. Wash them in water and dry with the help of filter paper. Separate the potato cylinders into 8 groups of 3 pieces each. Weigh each group separately and note down the weight. Now put each group into separate test tubes containing glucose solution of different concentrations mentioned. Also place one group into distilled water kept in another test tube. There will be seven dilutions of glucose solution and another will be distilled water. Keep the tubes for about 45 minutes. At the end of this period remove groups (3 pieces each) of potato cylinders from the test tubes, wipe them with filter paper and weigh them separately. Note down the loss, increase or no change in weight in each group. Tabulate the observations as follows

| No. | Contents | Initial Weight of potato cylinders | Final Weight of potato cylinder | Difference + or - |
|---|---|---|---|---|
| 1 | 0.1m | 4.5 gms | 4.9 | 0.4+ |
| 2 | 0.2m | 4.5 gms | 4.7 | 0.2+ |
| 3 | 0.25m | 4.5 gms | 4.5 | 0.0 |
| 4 | 0.3m | 4.5 gms | 4.4 | 0.1- |
| 5 | 0.35m | 4.5 gms | 4.3 | 0.2- |
| 6 | 0.4m | 4.5 gms | 4.2 | 0.3- |
| 7 | 0.5m | 4.5 gms | 4.1 | 0.4- |
| 8 | Distilled water | 4.5 gms | 5.0 | 0.5+ |

(The weights given above are arbitarary and serve only to illustrrate the example)

**Observations :** From the above it will be noticed that maximum increase in weight is noticed (0.5 gms) in the case of water, while there is slight increase in 0.1 m and 0.2 m, whereas in molarities above 0.3 there is decrease. In 0.25m however there is neither decrease nor increase. This concentration equals the concentration of the cell sap.

**Conclusions :** In concentrations below 0.25m there is entry of water into potato cylinders hence their weight increases. In concentrations above 0.2m there is exit of water from the potato cylinders hence their weight decreases. At 0.25m there is neither entry nor exit of water hence no change in weight. This concentration may be used to calculate the osmotic potential (OP) as follows.

OP = CRT

(Where C= molar concentration of the solution, R = gas constant equivalent to 0.082 and T=absolute temperature i.e. 273 + room tmperature Room temperature may be approximately 24°C)

Hence OP = 0.25 x 0.082 x 277

(This value will be a negative value as OP is always calculated as lack of so much pressure to accomodate certain quantity of water to bring its value to zero. When maximum water is present the pressure is zero).

### *23. Comparison of concentrations of two solutions in relation to cell sap by turrgidity method*

**Materials :** Fresh stem of any herbaceous plant such as *Synedrella*, *Amaranthus*, *Gynura* etc, Salt solution 1(weak) and 2 (strong), watch glasses.

**Procedure :** Cut the stem (internode) of the herbaceous plant into a cylinders of about 4 cms. Split into two halves logitudinally. Drop one half into salt solution 1 and the other into solution 2 taken in watch glasses. 2 leave them aside for about 30 minutes. Observe the changes in the stem.

**Observation :** The piece immersed in a weak solution (solution1) will bend outwards i.e. towards the epidermis and the piece put in strong solution (solution 2) will bend inwards i.e. towards the pith.

**Conclusion :** In a hypotonic solution (solution 1) the epidermal cells will not absorb water readily due to the presence of cuticle. The pith cells however absorb water (due to osmosis) readily and increase their size. As a result due to the streching of the pith cells the stem bends outwards. In a hypertonic solution (solution 2) the pith cells lose water much faster than the epidermal cells. As a result they (pith cells) shrink resulting in the inward curvature of the stem.

### *24. Development of turgor pressure due to osmosis*

**Materials :** Collodion sac, 10% sugar solution, water, beaker, thread, stand etc.

**Procedure :** Prepare a collodion sac as explained in experiment No. 19. Fill the sac with 10% sugar solution, twist the open end bend it back and tie it securely with the help of a thread. Immerse the sac in water contained in the beaker after securing it to a stand. Observe the changing of shape of the sac.

**Observation :** It is noticed that the collodion sac starts bulging due to the entry of water from the beaker.

**Conclusion :** The collodion membrane is selectively permeable. It allows water only but not solutes. Since the solvent concentration in the beaker is more, water enters into the collodion sac due to osmosis. As water enters, the sugar solution presses against the inner wall of the collodion sac. This pressure is called the turgor pressure. One to this pressure the collodion sac bulges.

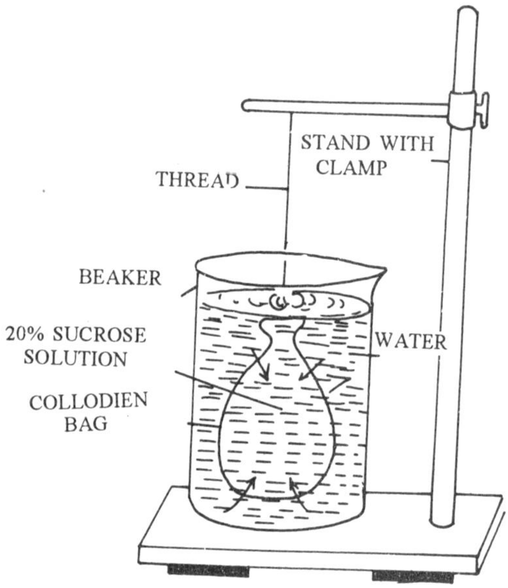


**Fig 14.10 Demonstration of Turgor Pressure**

***25. Demonstration of endosmosis and exosmosis Using boiled Eggs***

**Materials :** Wide mouth bottles, 10% salt solution, water, boiled eggs

**Procedure :** Take two boiled eggs and carefully remove the shell. Let the egg be intact without any damage. Select two wide mouth bottles (A and B) where the diametre of the mouth of the bottle should be such that, the deshelled egg can be squeezed through (the mouth) with some difficulty. Let bottle A have distilled water and bottle B have 10% salt solution. Squeeze one egg each into these bottles. Allow them to be immersed in the liquids for 30 minutes. At the end of this period try to remove the eggs from the bottles by inverting these bottles.

**Observation :** It will be very difficult to take out the egg from bottle A without breaking it as it would have swollen. From bottle B however, the egg comes out easily as it would have shrunken when compared with the original size.

**Conclusion :** The liquid (water) in bottle A is hypotonic when compared with the sap in the egg. The egg is surrounded by a selectively permeable membrane and it acts as an osmotic system and draws water into it. As a result the egg swells. This is called Endosmosis as water enters into a confined system (egg). In bottle B the liquied is a hypertonic solution whose concentration is more than that inside the egg. It is a weak solvent. Hence water goes out of the egg and into the bottle. As a result the egg shrinks in size and can easily come out of the bottle. This is called exosmosis as water comes out of a confined system (egg).

## Absorption of Water

***26. Absorption of water by roots***

**Materials :** A potometer, beaker, balsam plant, rubber tube, bent glass tube, measuring cylinder, knife etc.

**Procedure :** Select a potometer which has a reservoir fitted to a glass jar and a graduated capillary tube attached at another end (see figure). Pull out a healthy balsam plant from the soil with intact roots. Wash the root system under running water to remove all adhering soil particles. Insert the root system into the jar filled with water through a split cork. Now cut off the shoot system a little above the cork and connect the stump to a bent glass tube with the help of a rubber tube. Let the free end of the tube be inserted into a measuring cylinder. Carefully introduce an air bubble into the graduated tube and dip it into a beaker filled with water. Leave the apparatus to stand for some time. Observe the exudation of water into the measuring cylinder and absorption of water through the graduated tube.

**Observation :** The air bubble moves from one end of the graduated tube to another end as the water is absorbed by the root system. By opening the reservoir the air bubble can be pushed back and the movement of the air bubble can be monitored. The quantity of water absorbed by the plant can be calculated by measuring the distance travelled by the airbubble. The liquied that comes out of the stump gets collected in the measuring cylinder is due to root pressure as a result of water absorption. It may be noted that the amount of water collected in the measuring cylinder is approximatelly equivalent to the amount of water absorbed.

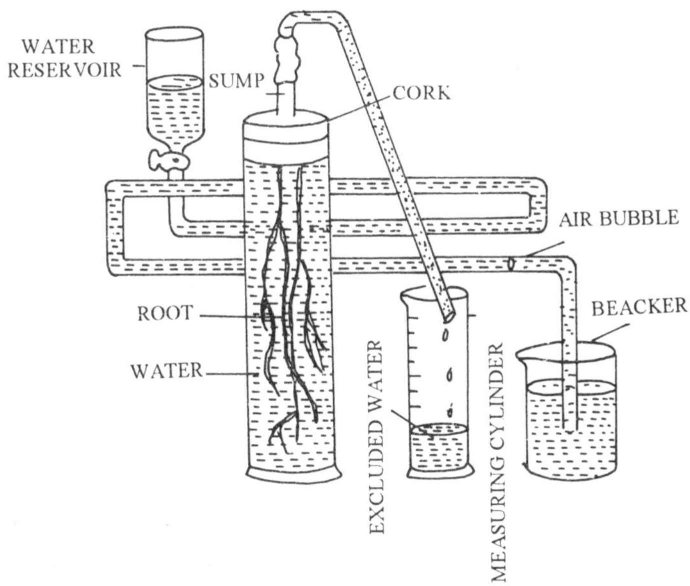


**Fig 14.11 Demonstration of Water Obsorption**

**Conclusion :** This experiment demonstrates the absorption of water through roots and its movement into the stem.

## ASCENT OF SAP

### *27. Root Pressure*

**Materials :** A healthy potted plant, rubber tube, blade or knife, mercury manometer, water, stand etc.

**Procedure :** Take a healthy potted plant. Keep it in sunlight. Water the plant well. After some time cut off the shoot a few inches above the soil level and attach a mercury manometer to it with the help of a rubber tube. Fill the tube below the bulb of manometer with water and above the bulb with mercury. Note the initial level of mercury. Keep the apparatus for a while. Note the level of mercury again.

**Observation :** There is increase in the level of mercury in the manometer.

**Conclusion :** Water being continuosly absorbed by the roots enters the cortext without much resistence. As the water reaches endodermis it does not have an easy passage and it has to get collected. But with the continued absorption of water and the unyielding endodermis a hydrostatic pressure is built up and as a result water is forced into the vasculature and it rises up to a certain level in the stem. This results in a rise in the level of mercury. This phenomenon of water rise in the stem due to the hydrostatic pressure built up by roots is known as root pressure.

***28. Demonstration of Path of Ascent of Sap***

**Materials :** Balsam plant eosin, water, beaker, slide, cover slip, microscope.

**Procedure :** Cut out the root portion of the balsam plant and keep it in a beaker containing water coloured with eosin. Leave it aside for about one hour. Observe the plant after that period. Cut a cross section of the stem and observe it under the microscope.

**Observation :** It will be seen that in the transparent stem of the balsam plant coloured lines or streaks appear indicating the path of upward movement of water. Transverse section of the stem observed under the microscope shows that only xylem vessels are coloured indicating that eosin coloured water moves up through them.

**Conclusion :** This experiment shows that ascent of sap in the stem is through the xylem elements

## TRANSPIRATION

***29. Demonstration of transpiration by bell Jar Method***

**Materials :** A healthy potted plant, glass plate, vaseline oil, cloth or polythene bag, thread etc.

**Procedure :** Select a healthy potted plant and keep it on a glass plate. Cover the top of the pot with a polythene bag or oil cloth leaving a hole for the plant. Tie the cloth securely to the pot. Cover the pot with a bell jar and apply vaseline to the edges of the bell on the glass plate to make it air tight. Now keep the entire set up in sun light. Observe the apparatus after about 30 minutes.

**Observation :** Small drops of water appear on the inner side of the bell jar.

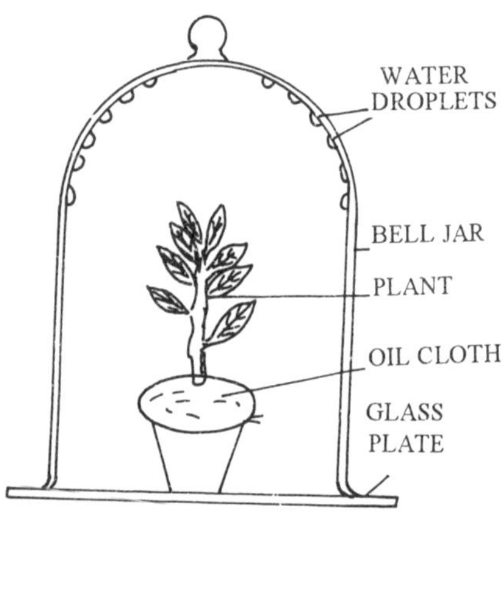


**Fig 14.12 Demonstration of Transpiration (Bell jar method)**

**Conclusion :** This experiment demonstrates transpiration. As the water vapouurs come out of the plant they get condensed on the walls of the bell jar and form water drops.

***30. Demonstration of unequal transpiration from the leaf surfaces by Cobalt chloride method***

**Materials :** A healthy potted plant with dorsivental leaves, filter paper, cobalt chloride solution (5%) slides, thread, desicator etc.

**Procedure :** Prepare a slightly acidic solution of cobalt chloride (5% solution). Dip pieces of filter paper in the $CoCl_2$ solution, squeeze the excess solution from the filter paper and dry in a desicator. Select an intact leaf in a healthy potted plant. With the help of a cotton wad clean both surfaces of the leaf. Take two pieces of cobalt chloride paper and place one piece on either surface of the leaf. Secure the $CoCl_2$ paper to the leaf by placing dry glass slides on them and tying them with the help of thread. Keep the entire set up in bright light and watch the change in colour of the filter papers on either side of the leaf.

**Observation :** Within few minutes the $cocl_2$ paper attached to lower epidermis which is blue in colour turns to pink whereas the $cocl_2$ paper attached to the upper epidermis turns pink only later.

**Conclusion :** Dry $Cocl_2$ paper will be blue in colour. They turn pink only in the presence of moisture. The paper on lower epidermis turns pink first because of more moisture output. Similarly the paper on upper epidermis turns pink later because of low moisture output. Moisture ouput in the form of vapours in due to transpiration. This experiment shows the rate of transpiration to be more in lower epidermis than in upper epidermis. Quantity of water vapours lost from an area due to transpiration is directly proportional to the number of stomata in that area. This experiment also demonstrates that the lower epidermis has more number of stomata per unit area than the upper epidermis.

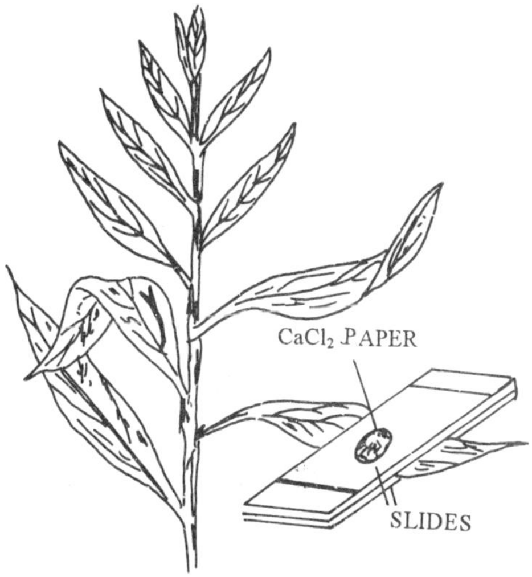


**Fig 14.13 Demonstration of Unequal Transpiration (Cobalt Chloride method)**

***31. Comparison of rates of transpiration between upper and lower epidermis of leaf by bell jar Method***

**Materials :** A healthy potted plant, two small bell jars, test tubes, thread, stand, mercury manometer, cork, desicator, grease, anhydrous $CaCl_2$, balance

**Procedure :** Select a healthy leaf in a well watered plant and place the open end of the bell jars on the upper and lower epidermis as shown in the figure. Fix the two bell jars to a stand. Now with the help of a split cork insert a mercury manometer to the other end of each bell jar. This is to disconnect the air supply between the exterior and interior. Carefully weigh about 6 grams of $CaCl_2$ kept in a desicator. Divide into two equal portions and eqaulise the weight of both the test tubes. Keep one test tube in the lower bell jar and hang another with the help of a thread in the upper bell jar. Keep the apparatus in sunlight for about 1 hour. At the end of this period remove the test tubes with $CaCl_2$ from both the bell jars and weigh them separately. Note down the increase or decrease in weight.

**Observation :** It is noticed that the weight of $CaCl_2$ tube in the lower bell jar has increased more than that of the tube in the upper bell jar.

**Conclusion :** The rate of transpiration is more in the lower epidermis than in the upper epidermis. As a result $CaCl_2$ kept in the lower bell jar absorbs more moisture and weighs more than the $CaCl_2$ tube kept in the upper bell jar. Higher rate of transpiration in the lower epidermis is due to more number of stomata per unit area than in the upper epidermis.

***32. Unequal Transpiration in the lower and Upper epidermis - 4 leaves experiment***

**Materials :** Four fresh dicotyledonous leaves, vaseline, thread, stand etc.

**Precedure :** Select four healthy dicotyledonous leaves of approximately same age from a single plant. Apply

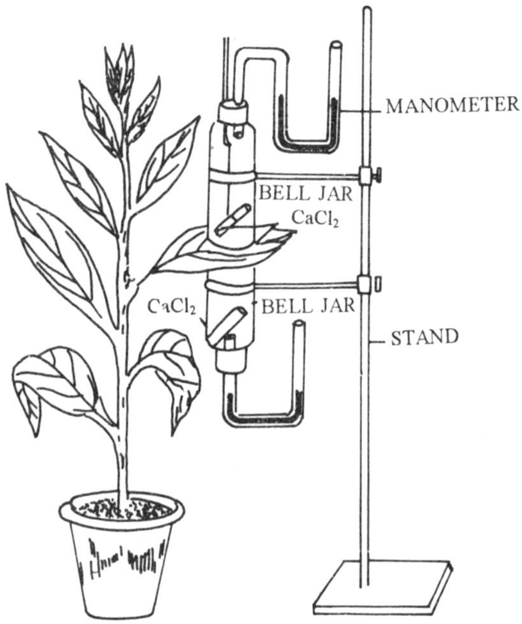


**Fig 14.14 Comparison of rate of Transpiration between upper and lower surface of leaf**

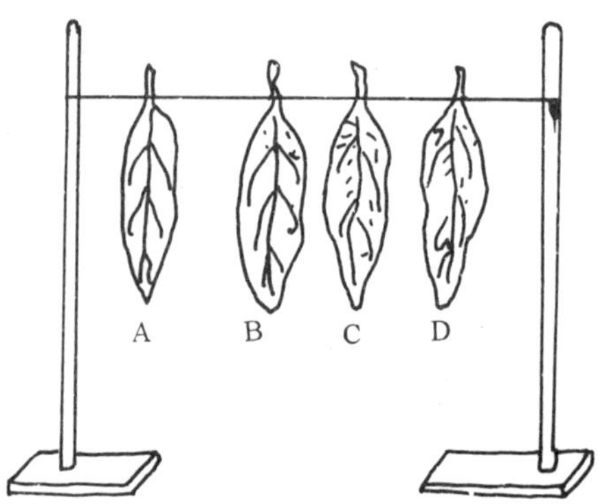


**Fig 14.15 Four leaves Experiment**

vaseline to the cut end of the petiole and hang them on to a thread streched between two stands by securely tying them. Smear vaseline selectively to the surfaces of the leaf as follows -

(a) Leaf 1 : Vaseline on both surfaces
(b) Leaf 2 : Vaseline on only upper surface
(c) Leaf 3 : Vaseline on only lower surface
(d) Leaf 4 : No vaseline (control)

Let the apparatus stand for about 2 hours

**Observation :** Leaf 4 wilts first followed by leaf 2 and 3, while leaf 1 does not wilt and remains fresh.

**Conclusions :** Leaf 4 wilts first because of unchecked transpiration from both upper and lower epidermis while leaf wilts next because of unchecked transpiration from lower epidermis. Leaf 3 takes a little longer because transpiration from the lower surface which is always high is checked because of smearing vaseline. Leaf 1 remains fresh for a longer time because transpiration from both upper and lower epidermis is checked. Since he leaf 2 wilts earlier than leaf 3 it is clear that the rate of transpiration is more in lower epidermis than in upper epidermis.

***33. Comparison of rate of absorption and rate of Tranpiration***

**Materials :** Wide mouthed bottle with a graduated side tube, oil, split cork, a healthy plant, balance.

**Procedure :** Select a wide mouthed bottle with a graduated side tube and fill it with water. Put a drop of oil on to the water in the side tube to check evaporation. Now pull out a healthy plant from soil with roots. Wash it thoroughly and fix it to the mouth of the bottle with the help of a split cork so that the root system is submerged in water. Weigh the entire apparatus and note down the weight. Similarly note down the level of water in the side tube.

Keep the apparatus in sunlight for about one hour. At the end of this period weigh the apparatus and also note the level of water in the side tube.

**Observation :** There would be reduction in the weight of the aparatus and at the same time the level of water would have come down in the graduated side tube.

**Conclusions :** Lowering of water level in the side tube is due to absorption of water by the plant while reduction in weight is due to loss of water by way of transpiration. The difference in the weight of the apparatus is equal to the amount of water absorbed by the plant. This shows that the amount of water absorbed is approximately equal to the amount of water transpired.

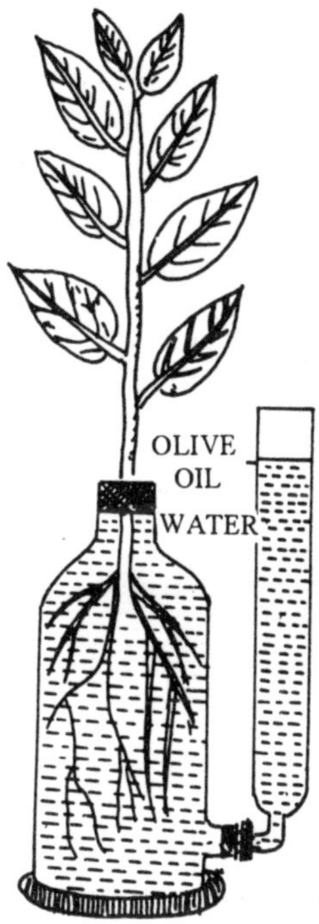


Fig 14.16 Apparatus to compare transpiration and Absorption

### *34. Development of suction force due to transpiration*

**Materials :** An open ended glass jar with a side tube, a healthy plant, stand, beaker, mercury, water, glass tube etc.

**Procedure :** Take a glass jar with a side tube. Fill in water and close all the three ends with the help of corks. To the lower end of the jar fix a capillary glass tube and dip the other end of the tube in a beaker containing mercury. Initially the capillary tube is also filled with water. Fix a plant with roots into the side tube with the help of a split cork. Make the apparatus air tight and keep it in sun light. Observe after about 45 minutes.

**Observation :** It is observed that the level of mercury rises in the capillary tube.

**Conclusion :** As water evaporates due to transpiration a suction force develops and water is made to enter the plant. This results in the rise in the level of mercury in the capillary tube. Suction force can be calculated as follows.

$Sf = \pi r^2 \times 13.6$ where

sf = suction force

13.6 is the relative density of mercury and value of $\pi$ is 22/7.

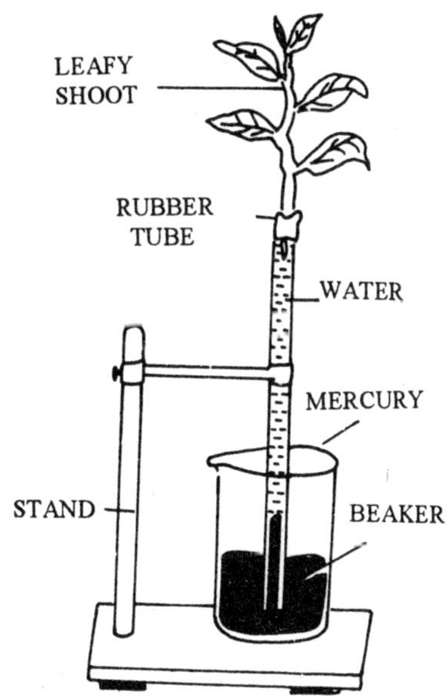


Fig 14.17 Development of suction force due to transpiration

*35. Measurement of rate of transpiration by Ganong's potometer*

**Materials :** Ganongs potometer, beaker, plant, eosin water.

**Procedure :** The apparatus consists of a horizontal graduated tube which is bent in opposite directions at either ends. One bent end is wide and the other end is narrow. A reservoir is fixed to the horizontal tube near the wider end. The reservoir has a stop cock to regulate the water flow. The apparatus is filled with water by opening the reservoir. A twig or a small plant is fixed to the wider arm through a split cork. The othe bent end of the horizontal tube is dipped into a beaker containing water coloured with eosin. An air bubble is introduced into the graduated tube at the narrow end (arm). As transpiration takes place, the air bubble will move from one end of the graduated tube to the other. After it has reached the reservoir, it (reservoir) is opened to push the air bubble back and then its movement can be watched again.

**Observation :** Record the time that is needed for the air bubble to move from one end to the other; make many observations to obtain the average of time taken to travel the distance.

**Conclusion :** The Ganong's potometer measures the rate of transpiration indirectly by measuring the rate of absorption. The rate of transpiration is approximately equal to the rate of absorption. By using the following formula the amount of water transpired may be caluclated.

$V = \pi r^2 l$

Where V = value of water, $\pi$ = 22/7, $r$ = radius of graduated tube, 1 = distance travelled by the air bubble.

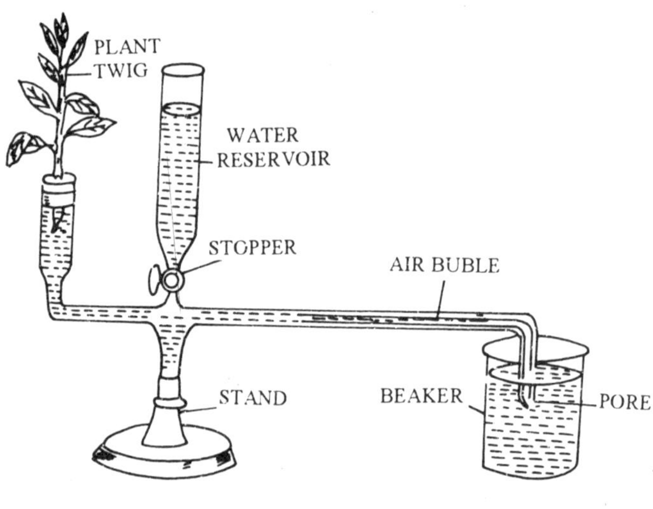


**Fig 14.18 Ganong's Potometer**

*36. Measurement of Transpiration by Farmers' Potometer*

**Materials :** Farmer's potometer, beaker, water, vaseline, plant

**Procedure :** The Farmer's potometer consists of a wide mouthed bottle. A cork with three holes is fitted on to the mouth of the bottle. One whole is fixed with a thrice bent tube with a scale fixed on to it horizontally, the second hole has a reservoir with a stopper fitted onto it and through the third hole fix a twig under water.

Fill the apparatus with water, open the reservoir so that the tube gets filled with water. Make the apparatus air tight by applying vaseline. Dip the free bent end of the horizontal tube into a beaker containing water. Introduce an air bubble into the tube. Keep the apparatus in sunlight for some time.

*Observation* and *Conclusion* : Same as in experiment No. 35.

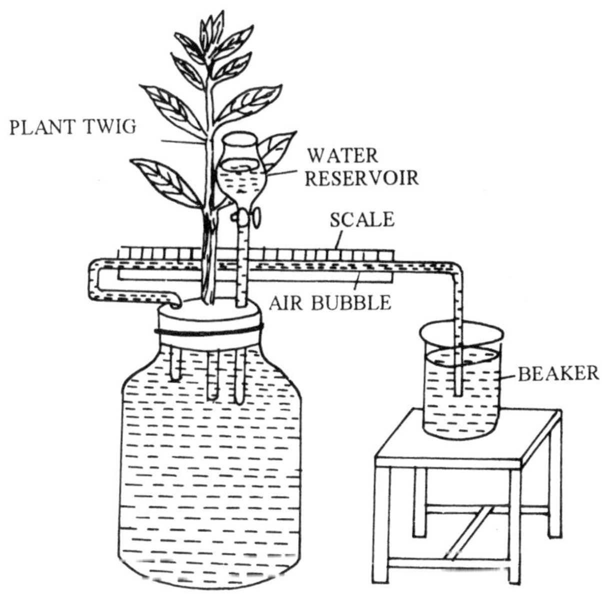


**Fig 14.19 Farmer's Patometer**

***37. Demonstration of Continuity of air spaces in the leaf***

**Materials :** Wide mouthed bottle, cork with two holes, aspirator, bent glass tube, vaseline, a leaf with a long petiole such as *Colocasia*, lotus etc.

**Procedure :** Take a wide mouthed bottle which is half filled with water. Close the mouth with a two holed cork. Introduce a leaf into the bottle through one hole so that the petiole is dipped in water. Through the other hole fix a bent tube so that the tip of the tube shoudl be few centimetres above the water level. Connect the other end of the tube to an aspirator. Apply vaseline to all connecting points so that the apparatus is air tight. Start the aspirator and observe the cut end of the petiole.

**Observations :** Air bubbles start coming out of the petiole as long as the aspirator is on.

**Conclusion :** As the aspirator sucks air the suction is transmitted through the petiols along the intercellular spaces and up to stomata. Stomata suck air and it travels down the petiole and released into water. This is possible only when the air space is continious within the plant.

*38. Demonstration of Guttation*

**Materials :** *Nasturtium* plant in a pot, glass plate, bell jar, plastic sheet, vaseline, thread.

**Procedure :** Keep the potted planted (*Nasturtium* on a glass; plate cover the pot with a plastic sheet with a hole to allow the plant to be exposed. Invert the bell jar on the pot. Let the mouth of the bell jar at the narrow end be covered with a single holed cork. Fix a bent tube through this whole so that one end is in the bell jar while the other end is fixed to an aspirator. Make the apparatus air tight by applying vaseline. Keep the apparatus in sunlight and switch on the aspirator and observe.

**Observation :** As the air inside is sucked by the aspirator water drops appear on the margins of leaves.

**Conclusion :** As the air is sucked, hydathodes present in the leaves push out water due to the sucking force. Water here comes out physically and it is known as guttation. In transpiration however water comes out in the form of vapours and it does not settle on the leaves, but gets condensed on the bell jar.

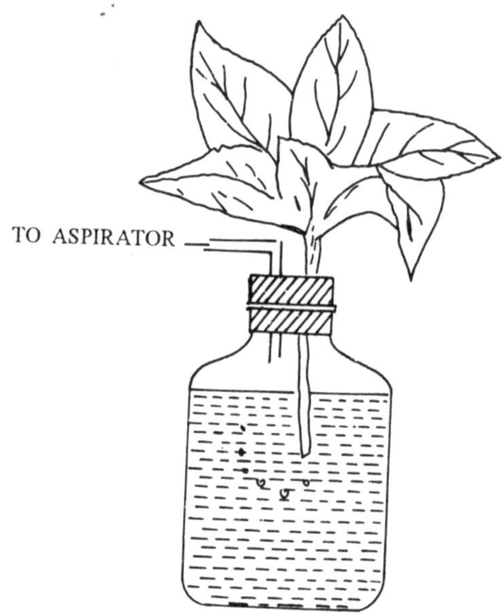


**Fig 14.20 Demonstration of Continuity of Air Spaces in Plant**

*39. Measurement of Stomatal Opening by Darwin's Potometer*

**Materials :** Darwin's prometer, A healthy potted plant, water, clips, stopwatch, beaker, adhesive etc.

**Procedure :** Darwin's prometer has a 'T' shaped tube. Attach a rubber tubing to both the arms of 'T'. Close one of the rubber tubings with the help of a clip while to another attach a small cup. Fix the cup to the lower surface of a leaf of a potted plant with an adhesive. Smear vaseline to the edges of the cup to make it air tight. Dip the stem of 'T' in a beaker containing water open the clip and suck water through it. Water column rises in the stem of 'T' tube. Note the level. Separate the leaf from the plant and after sometime note the level of water in 'T'.

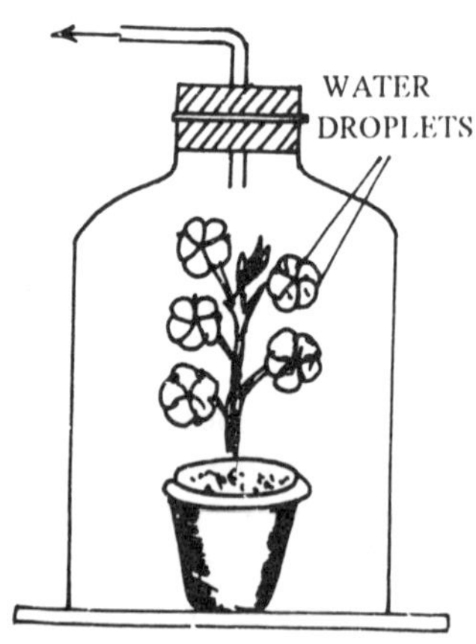


**Fig 14.21 Demonstration of Guttation**

**Observation :** When the leaf is attached to the plant the level of water in 'T' goes down gradually while the level remains unaffected if the leaf is detached from the plant.

**Conclusions :** When air is sucked through the rubber tubing its pressure is lowered in the potometer. In the intact leaf stomata are open and transpire, with the result water vapours enter the cup and push down the water column. When the leaf is detached stomata remain closed, transpiration does not occur and water level in the 'T' tube remains same.

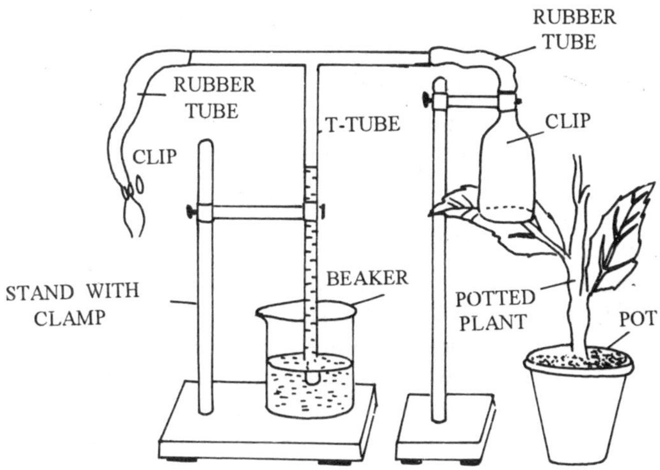


**Fig 14.22 Demonstration of Stomatal opening by Darwin's potometer method**

## PHOTOSYNTHESIS

### *40. Evolution of $O_2$ during Photosynthesis*

**Materials :** Beaker, watch glass, funnel, test tube, *Elodea* or *Hydrilla* plant, Pyrogallol.

**Procedure :** Take a glass beaker half filled with water and put some pieces of *Hydrilla* or *Elodea* plant into it. Invert a glass funnel over it. Fill a test tube with water and carefully invert it over the stem of the funnel so that the mouth of the test tube is under water. Keep the apparatus in sunlight and observe

**Observation :** Small air bubbles appear from the cut ends of braches of the plants and they enter the test tube. Gradually water in the test tube gets downwardly displaced and it (test tube) becomes filled with the gas. The gas can be tested by one of the following methods.

(a) cover the mouth of the test tube and carefully take out the test tube. Introduce a burning splinter into the test tube. It glows brightly.

(b) Dissolve some amount of pyrogallol in water. The gas is dissolved in pyrogallol and the level of the water in the test tube rises again.

**Conclusion :** Both the tests performed above will show that the gass is oxygen. The plants kept in the beaker photosynthesize in the presence of sunlight and release oxygen which comes out of the plant in the form of air bubble.

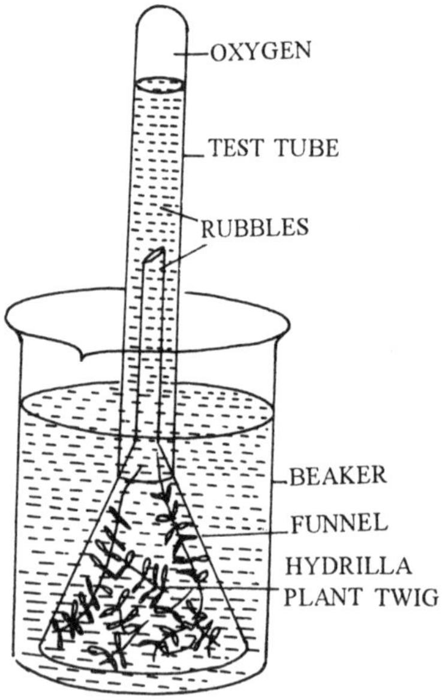


**Fig 14.23 Evolution of $0_2$ during photosynthesis**

***41. Chlorophyll is necessary for photosynthresis***

**Materials :** Plant with variegated leaves such as *Codeum, Coleus, Manihot* etc, butter paper, pencil 70% alcohol, potassium iodide solution, water bath

**Procedure :** Select a potted plant with variegated leaves and keep it in darkness for 72 hours so that it gets destarched. Bring out the plant and place it in bright sunlight. Select a leaf with variesated colours and placing a butter paper on the leaf trace the outline of various colours with the help of a pencil. After few hours pull out the leaf from the plant, boil it in 70% alcohol in a water bath so that the leaf gets decolourized. Pour KI solution on the leaf. Observe the change in colour.

**Observation :** Blue colour appears only in those areas which were green. This can be checked in the outline diagram of the leaf.

**Conclusion :** Chlorophyll pigment is necessary for photosynthesis. Only the green areas in the leaf photosynthesize and starch is present in those regions. That is why when KI solution is poured they turn blue (indication of starch). Other areas (non green) do not photosynthesize, do not have starch and do not turn blue on contact with KI solution.

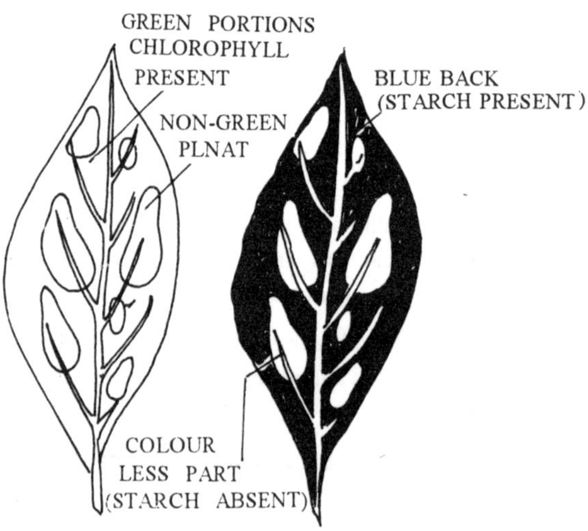


**Fig 14.24 Chlorophyll is necessary for Photosynthesis**

***42. Light is necessary for Photosynthesis (Ganong's light screen experiment)***

**Materials :** A potted plant, Ganong's light screen, beaker, 7% alcohol, KI solution, test tube, water bath.

**Procedure :** Keep a potted plant in darkness for about 72 hours so that it is starch free.

Fix a ganong's light screen to one of the leaves and place the pot in bright sunlight. After few hours pluck the leaf from the plant, remove the light screen and decolourize it by boiling (in a test tube having 70% alcohol) it in a water bath. Take out the decolourized leav and pour KI solution.

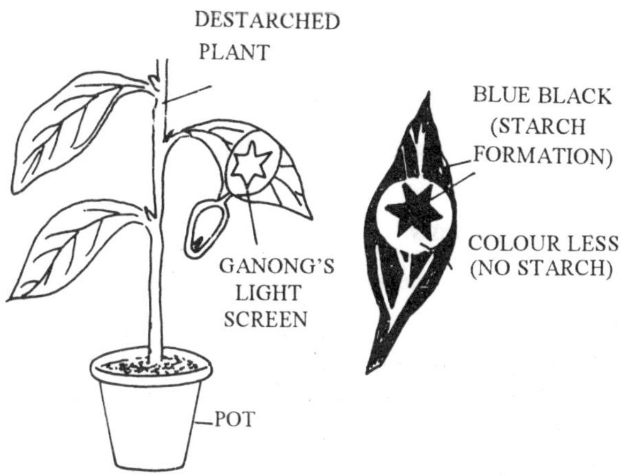


**Observation :** The area of leaf covered by light screen remains bleached while the other areas turn blue on contact with KI solution.

**Conclusion :** The area covered by the light screen does not get light hence starch is not manufactured in that area. As a result it does not turn blue on contact with KI solution. Other areas get light so they manufacture starch and turn blue on contact with KI solution.

**Fig 14.25 Light is necessary for photosynthesis**

***43. $CO_2$ is necessary for photosynthesis (Mohl's Half leaf experiment)***

**Materials :** Wide mouthed bottle, split cork, KOH, Healthy plant.

**Procedure :** Keep a potted plant in darkness for about 72 hours to remove starch. Select a leaf from such a plant and without removing it insert half a portion into a bottle through a split cork. Before inserting the leaf pour some quantity of KoH liquid into the bottle. Make the apparatus air tight by smearing vaseline to the edges of the cork. Keep the apparatus in sunlight for about 2 hours. At the end of the period remove the leaf from the bottle detach it from the plant and boil it is 70% alcohol to decolourize it. Pour Iodine solution and observe the change in colour.

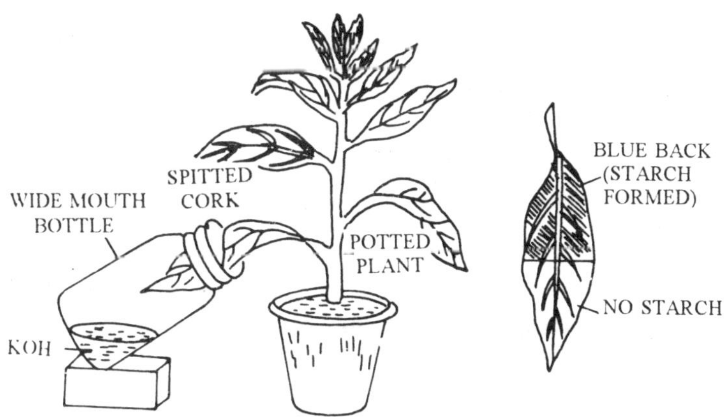


**Fig 14.26 Mohl's half leaf experiment**

**Observation :** The part of the leaf inside the bottle will remain bleached whereas the portion outside turns blue.

**Conclusion :** The portion of leaf inside bottle had all the necessary factors for photosynthesis except $CO_2$ because it was absorbed by KoH. The portion of leaf outside the bottle turns blue as it has starch manufactured by photosynthesis. It was provided with all the factors including $CO_2$. This demonstrates clearing that $CO_2$ is necessary for Photosynthesis.

### *44. Effect of different wave lengths of light on photosynthesis*

**Materials :** A healthy potted plant, Ganong's light box, beaker, 70% alcohol, KI solution, test tube, water bath.

**Procedure :** Ganong's light box is a small rectangular box with removable lid which accomodates glass plates of different colours. Select a leaf from a healthy potted plant which has been destarched (by keeping it in darkness for 72 hours) and insert it into Ganong's light box. Let the box have three colour plates viz. red (660 μm), blue (455 μm) and green (530 μm). Keep the apparatus in sunlight with this arrangement the leaf will be exposed to three wave lengths of light viz red, blue and green of the visible spectrum. After about 2-3 hours remove the leaf from light box detach it from the plant, decolourize it by boiling it in 70% alcohol in a water bath. Pour Iodine solution over the leaf and observe the change in colour.

**Observation :** Three portions of the leaf were exposed to three different rays of light. The segment of leaf exposed to red light turns deep blue, the part of leaf exposed to blue rays turns light blue and the part of leaf exposed to green light remians almost unchanged

**Conclusion :** Part of the leaf exposed to red rays turning deep blue indicates concentration of starch, light blue colour in the region exposed to blue rays is an indication of less starch and the region exposed to green light remaining unchanged indicates lack of starch. This experiment shows that for photosynthesis red and blue rays are necessary while green rays are not needed.

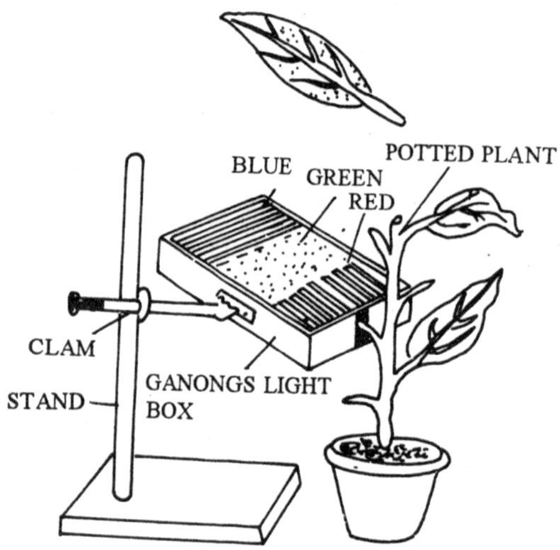


**Fig 14.27 Effect of different wavelength of light on Photosynthesis**

### *45. Measurement of Rate of Photosynthesis by Wilmott's Bubbler*

**Materials :** Wilmott's bubbler, *Hydrilla* or *Elodea* Plant, pond water, blade, stop watch etc.

**Procedure :** Take a Willmott's bubbler apparatus (It is a wide mouthed bottle with a single holed cork. A glass tube is inserted into the bottle while the upper end of the tube forms a hooked nozzle inside a broad tube called reservoir).

Fill the willmotts bubler with pond water. Select a branch of *Elodea* or *Hydrilla*, cut the shoot under water and place the cut end inside the tube fitted in the cork (see figure). Dip the other end of the plant in the bottle and close the stopper. Now pour some water into the reservoir above the bottle so that the nozzle is under water. Apply vaseline to the cork and keep the apparatus in sunlight and observe.

**Observation :** Small air bubbles escape from the cut end of the stem travel through the glass tube and finally come out through the nozzle. These bubbles which are uniform in size can be counted per unit time by using a stop watch

**Conclusion :** Release of air bubbles ($O_2$) from the cut end of the plant indicates the rate of photosynthesis. The number of bubbles per unit time can be calculated to measure the rate of photsynthesis. This is only an arbitary measurement.

***46. Demonstration of presence of starch in chloroplasts***

**Materials :** Filaments of *Spirogyra* or leaves of *Elodea*, Chloral hydrate Iodine solution, slide, cover glass, microscope etc.

**Procedure :** Mount a filament of *Spirogyra* a green alga under the microscope. Note down the shape of chloroplasts in the cells. Stain the cells with chloral hydrate Iodine solution and observe under the microscope.

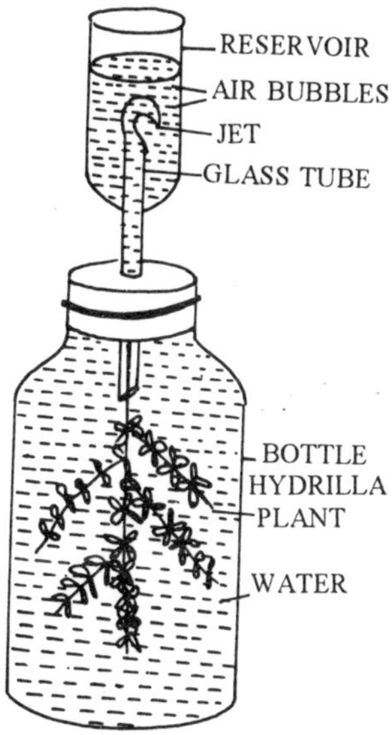


**Fig 14.28 Wilmotl's bubble experiment**

**Observation:** The chloroplasts in *Spirogyra* filament are spiral in shape. After staining the entire spiral band turns blue black. The chloroplasts turn blue black due to the presence of starch.

***47. Study of photosynthesis by gass analysis (Ganong's Photosynthometer)***

**Materials :** Photosynthometer, leaves, test tube, caustic potash, carbon dioxide generator.

**Procedure :** Ganong's photosynthometer consists of three parts - (1) a glass bulb A(2) a graduated tube C and (3) a connecting link B provided with a stop cock. B fits into the neck of A at the lower end and at the upper end the lower end of tube C fits into it. The total volume of the whole apparatus is approximately 103 ml. There are two holes a and b in A and B respectively so that they are at the same level when A fits into B.

Place a few fresh green leaves in the bulb of 'A' together with a little quantity of water. Now invert the graduated tube 'C' close its stop cock and fill it with 5 ml of water. Fix 'C' on to 'B' and close the stoplock of latter and fill water into lower half portion of 'B' and carefully covering it by hand dip into a beaker with water. Clamp the tube so that water in the beaker is on level with hole b. Open the stop cock of B as well as of 'C' and connect 'C' to a carbon dioxide generator. Stop the $Co_2$ entry into tube 'C' as soon as the water in tube 'C' falls to the level of water in beaker. Tube 'C' now has 5 ml of $Co_2$ (by displacing 5 ml of water). Close both the stop cocks ('C' and 'B\) and fit the whole thing ('B' and 'C') on to the bulb 'A'. Now the openings a and b (in 'A' and 'B' parts) are on the same level. this will equalise the pressure inside the bulb A to the atmospheric level. Twist 'B' slightly now so that a and b move apart and the connection becomes airtight. Place the apparatus in sunlight and open the stopcock of 'B' to allow $CO_2$ into bulb A (with leaves). Allow the apparatus to stand for 6 - 8 hours. At the end of this period, remove 'B' and 'C' together from 'A' ande dip them in water. Separate 'B' from 'C' underwater and open the stop cock of 'C' so that water enters into it up to zero mark. Fill a test tube with caustic potash and connect it to 'C' near the stop lock with the help of a rubber tube provided with a pinch cock. Lift the graduated tube ('C') from water by closing the lower end with hand and open the pinch cock as well as the stop cock to allow potash fluid to flow into 'C'. Shake thoroughly and drain the potash back to the tube and close the pinch cock as well as the stop cock. Now, still closing the lower end of 'C' with hand dip it into water, disconnect the test tube with caustic potash. Place the tube (in water) in such a way that the zero mark is on level with water. Take out the hand and water will rise into the tube up to a mark.

**Observation and Conclusion :** The amount of water which rises up in the tube is equal to the amount of $CO_2$ used up for photosynthesis. Originally 5 ml of $CO_2$ is present. Now if water rises up to 2 ml, it means 2 ml of $CO_2$ is used up for photosynthesis and 3 ml is left unused. (It is possible that this may not give an accurate amount of $CO_2$ used in photosynthesis, because $CO_2$ is slightly soluble in water).

***48. Extraction of Chlorophyll Pigments By Solvent wash method***

**Materials :** Fresh leaves (spinach, tecoma etc), or 3 grams of dry leaf powder, petroleum ether, Acetone, methyl alcohol,methanolic KOH, Separating funnel, stand, beaker, pestle, mortar etc.

**Procedure :** Take 5 gms of fresh leaves of spinach or *Tecoma.* Wash them thoroughly. Wipe off moisture with the help of a filter paper. Place them in a pestle and mortar and grind with 50 ml of ice cold acetone. Collect the liquid and filter. Discard the residue and collect the filterate. The filterate (acetone extract) consists of chlorophyll a, chlorphyll b, *carotene* and *xanthophyll.*

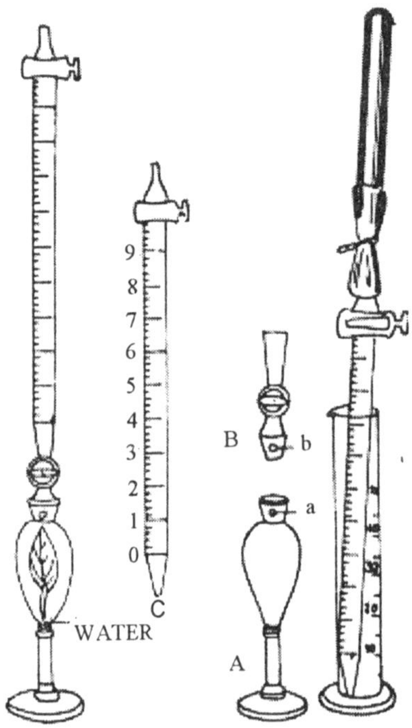


**Fig 14.29Ganong's Photosynthometer**

Instead of fresh leaves, leaf powder may be used. Collect fresh leaves and dry them in an oven. Powder the dry leaves. Weigh 3 gms and soak it in 40 ml of 80% acetone containing a pinch of CaCo3. Stir the solution and filter. The filterate acetone extract consists of all the pigaments mentioned above.

Add 60 ml of petroleum ether to 40 ml of acetone extract in a separateing funnel. Gently rotate the funnel. Slowly add 75 ml of distilled water along the sides of the funnel to avoid emulsification. Rotate the funnel again gently and fix it to a stand. Gradually two layers of the liquid are formed. Of these the lower is acetone - water layer. This may be discarded. The upper layer (petroleum ether layer) consists of all the pigments. Add 50 ml of distilled water gently to the petroleum ether layer. Again two layers are formed. Discard the lower layer. The upper layer may be washed again with 50 ml of distilled water and discarding the lower layer.

Add to the petroleum ether layer 50 ml of 92 % methyl alcohol. Gently rotate the separating funnel. Allow it to stand. Tow layers are formed viz. the lower methyl alcohol layer and the upper petroleum ether layer. Draw the two layers into separate beakers. Lower layer consists of chlorophyll b and xanthophyll. The upper layer consists of chlorophyll a and carotene. Further procedure for two layers is as follows -

1. **Upper Petroleum ether layer :** Take 30 ml of this layer and to it add 15 ml of freshly prepared 30% methyl alcoholic potassium hydroxide solution in a separating funnel. Rotate gently. Two layers are formed. Add 30 ml water and shake. The upper petroleum ether layer consists of carotene while the lower methanolic KoH layer consists of chlorophyll a. Collect these two layers in separate tubes.

2. **Lower Methyl Alcohol Layer :** Take 50 ml of methyl alcohol layer in a separating funnel and add 50 ml of diethyl ehter. Mix them thoroughly by rotating the funnel. Two layers are formed. Discard the lower methyl alcohol layer. The upper layer has thepigments. Add 15 ml of methanolic KOH to the upper layer. Shake gently for about 10 minutes. Add 30 ml of distilled water and shake again. Two layers are formed. The lower methanolic KoH fraction consists of chlorophyll b, while the upper diethyl either layer consists of xanthophyll. Collect them separately.

**Observation and Conclusion :** The four pigments separated can be characterised by checking their peak absorption spectra. The following is a flow chart for the process given above.

Leaves of Spinach, Tecoma etc
(5 gms)
↓ Grind in 50 ml of ice cold acetone and filter
Leaf extract
↓ Add 60 ml of petroleum ether and rotate gently
Separating funnel
↓
Wash gently with 75 ml of distilled water
↓
Two layers of the liquid
↓

- Upper petroleum ether layer
  ↓
  Wash with distilled water three times and discard lower layer every time
  ↓
  Add 50 ml of 92% methyl alcohol
  ↓ Roate gently
  Two layers
  ↓
  - Upper petroleumether layer
    ↓
    Add 15 ml of methanolic KoH and rotate gently
    ↓
    Shake for 10 minutes
    ↓
    Add 30 ml of distilled water and shake
    ↓
    Two layers
    ↓
    - Petroleum ether layer (**Carotene**)
    - Methanolic KOH layer (Chlorophyll a)
  - Lower methyl alcohol layer. Collect in alcohol funnel
    ↓
    Add 50 ml of diethyl ehter
    ↓
    ↓
    Wash with 30 ml of water
    ↓
    Two layers
    - Upper ether layer
      ↓
    - Lower methyl alcohol layer Discard
- Lower acetone - water layer discard

Add 15 ml of methanolic KoH
(30% and Shake
↓
Add 30 ml water and shake
↓
Two layers
↓

| ↓ | ↓ |
|---|---|
| Upper diethyll ether layer (**xanthophyll**) | Lower Methanolic Koh Layer (**Chlorophyll b**) |

***49. Separation of leaf pigments by circular paper Chromotography***

**Materials :** Two large petridishes, Whatman No. 1 filter paper, acetone, chlorophyll extract.

**Procedure :** Take a circular piece of whatman No. 1 filter paper and apply the chlorophyll extract to the centre of the paper in the form of a spot. Be careful not to allow the spot to spread. Use a hair dryer if necessary to dry the spot. Cut the filter paper from the periphery to the centre in the form of a small strip to form a 'wick'. Now place the filter paper over a small petridish (diameter must be smaller than that of the filter paper) containing about 30 ml of 20% acetone. Bend the wick in such a way that it dips in the acetone. Cover a large petridsh over the paper to prevent evaporation. Keep it for about 30 minutes. Watch the progress of the solvent. Remove the filter paper before the solvent reaches the outer periphery (of the paper). Dry it and observev the rings that appear. Measure the distance of the rings from the centre.

**Observation and Conclusion :** Four rings of coloured streaks appear. Of these the one closest to the centre of the filter paper is chlorophyll b, followed by chlorophyll a, xanthophyll and carotene. Rf values for each ring may be calculated as follows.

$$Rf = \frac{\text{Distance travelled by ring of pigment}}{\text{Distance travelled by the solvent}}$$

Rf value will be always less than 1 as distance travelled by the solvent is more than the distance travelled by the compound.

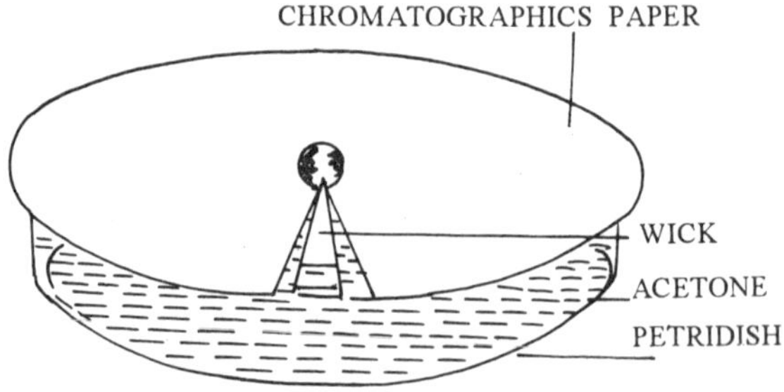


**Fig 14.30 Circular paper chromatography for separation of leaf pigments**

### *50. Separation of leaf pigments by Ascescending paper Chromatography*

**Materials :** Cylindrical or rectangular jar with a tight fitting lid, chlorophyll extract, whatman No.1. Filter paper, petroleum ether, acetone, hook or thread, micro pipette, pencil etc.

**Procedure :** Prepare the chlorophyll extract as explained in experiment 48. Cut out a long strip of whatman No. 1 filter paper and draw across a pencil line about 2-3 centimeters from the end. With the help of a micropipetle draw a little quantity of concentrated chlorophyll extract and spot it on the centre of the pencil line. Allow the spot to dry (use a hair dryer if necessary). Apply some more quantity of the extract to the same region. Be careful not to allow the spot to expand on the paper.

Now select a jar of suitable size (longer than the filter paper strip) with a tight fitting split cork. Pour 50 ml of the solvent (petroleum ether : acetone -= 100 : 12) into the jar. Fix the upper end of the filter pape (i.e away from the spot region) to the split cork and insert the paper into the jar so that the spot end dips into the solvent below. The paper should be dipped in such a way that the spot is above the level of the solvent. Apply vaseline to the edges of the cork. Allow it to settle for about 30 - 45 minutes. Keep observing.

**Observation :** The solvent migrates up along the paper strip and carries the spot along with it. Gradually 4 different coloured spots apper at different distances from the original spot. Remove the paper from the jar before the solvent reaches the upper edge of the pape and dry it.

**Conclusion :** The four spots that appear are (starting from a distance closest to the original spot) . Chlorphyll b, chlorophyll a, xathophyll and carotene. Calculate the Rf values as given in experiment 49.

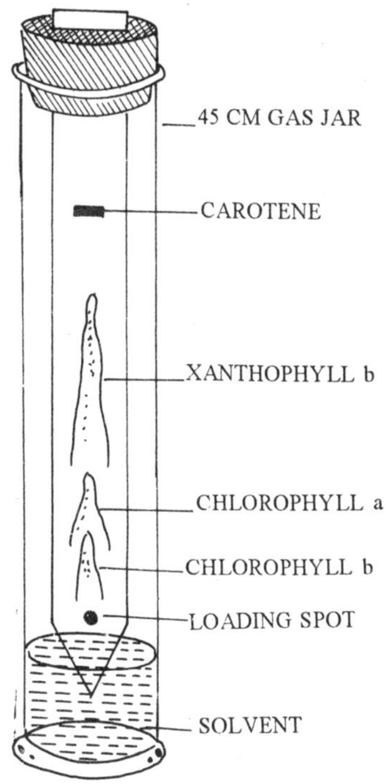


**Fig 14.31 Ascending paper chromatography for separation of leaf pigments**

### *51. Demonstration of Hill's reaction in isolated Chloroplasts (Dye reduction method)*

**Materials :** Spinach leaves, Sodium phosphate buffer (Nacl solution 0.25 ml, 0.1m potassium dihydrogen phosphate), 0.1%, 2-6 dichlorophenol indophenol (DCPIP), Ice bath, water bath, pestle and mortar, cheese cloth, test tubes, aluminium foil etc.

**Procedure :** There are two major steps (1) preparation of chloroplast suspension and (2) reduction of the dye DCPIP.

***(1) Preparation of Chloroplast Suspension :*** Prepare an ice bath and cool all the reagents, chemicals and glass ware used in the experiment. Weigh about 10 grams of fresh spinach leaves, wash them thoroughly and cut them into pieces. Remove, petiole, midrib and other hard materials and collect only the lamina portion. Macerate this tissue in pestle and mortar kept in an ice bucket. Alternatively the tissue can be homogenised in a waring blender with the operations conducted in a cold room.

Add 20 ml of sodium phosphate buffer at pH 6.5 (mix 0.24 m Nacl and 0.1m $KH_2 Po_4$ in the ratio of 25 : 10 ml). Filter the suspension with two layers of cheese cloth. Discard the residue and collect the filterate. Spin the filterate in a cooling centrifuge at 3000 rpm for 15 minutes. Discard the super natant and suspend the pellet in buffer and spin again. Discard the supernantant and resuspend the pellet in buffer. Stir it and transfer it to a tube kept in ice bath. This is the suspension of isolated chlorplasts.

*(2)* ***Reduction of Dye :*** Take two test tubes and add ingredients as follows :

A. 5.0 ml buffer + 0.6 ml distilled water + 2.0 ml of chloroplast suspension

B. Same as above.

Cover tube 'A' with aluminium foil or black paper so that no light can reach the chloroplast suspension. Let tube B be uncovered. To both 'A' and 'B' add a few drops of DCPIP. Keep both the tubes in water bath at room temperature and subject them to strong illumination.

After some time observe the contents of both tubes.

**Observation :** It is seen that the blue colour of the dye remains unchanged in the tube covered with black paper while in the tube exposed to light the dye would be decolourized.

**Conclusions :** During light reaction of photosynthesis water splits into H and OH. While OH recombines to form water, H is accepted by a hydrogen acceptor (NADP) which gets reduced. In the present experiment DCPIP is used instead of NADP. Chloroplasts receive light and water splits to form H and OH. DCPIP accepts H, gets reduced and becomes colourless

DCPIP + $2H_2O$ DCPIP (reduced + $O_2$

This experiment demonstrates that isolated chloroplasts can carry out photosynthesis even in a cell free medium.

## RESPIRATION

***52. Demonstration of Anaerobic respiration***

**Materials :** Petridish, test tube, germinating seeds, mercury, KOH crystals, forceps etc.

**Procedure :** Fill a test tube with mercury and invert it over a petridish also containing mercury. Now lift the test tube lightly and introduce few germinating seeds (gram, pea etc with testa removed) with the help of a forceps. The seeds will rise to the top of the test tube. Keep the test tube back in its position and allow the apparatus to stand for about 45 minutes.

**Observation :** A gas starts building up at the test tube and it pushes the mercury down. Now introduce a piece of caustic potash. This absorbs the gas and the level of mercury goes up and fills the test tube.

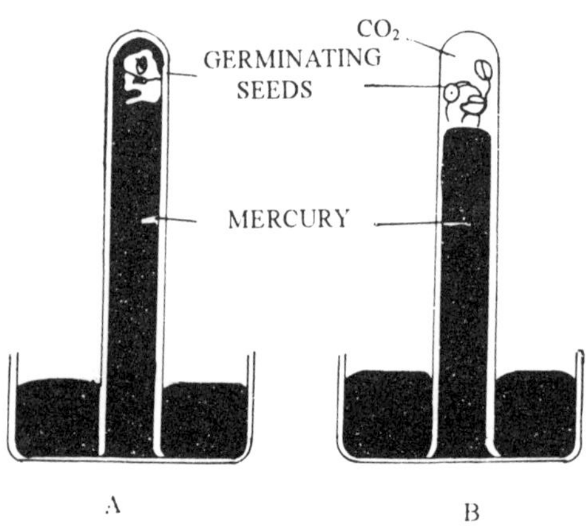


**Fig 14.32 Demonstration of Anaerobic Respiration**

**Conclusion :** Germinating seeds have a high rate of respiration. As they are kept in mercury they do not get oxygen and hence respire anaerobically. Since introduction of KOH absorbs the released gas it (gas) must be $CO_2$. This experiment demonstrates that $CO_2$ is released during anaerobic respiration

### *53. Demonstration of Fermentation (Kuhne's fermentation flask)*

**Materials :** Kuhne's flask, yeast extract or toddy, 5% glucose solution, KOH pellets

**Procedure :** Kuhne's flask consists of a slightly slanted glass tube with a side bulb attached to it. It has a basal stand. The standed tube has no opening at the top while the side tube with a bulb has an opening which can be closed with a stopper.

Fill the Kuhne's flask with 5% sugar solution in such a way that the slanted tube is full but the bulb is only filled up to half its capacity. Add yeast extract or few drops of toddy into the glucose solution. Close the opening of the bulb with a stopper or with cotton. Let the apparatus stand for some time.

**Observation and Conclusion :** Small bubbles of gas start rising in the slanted tube and push the level of the solution below with the result a gas filled space is created at the top. If this is allowed to continue the glucose solution fills the side bulb and it may overflow if there is no stopper. Introduce a piece of KOH into the tube through the bulb so that it reaches the gas filled top, the gas is absorbed and the glucose solution once again moves to the top of the tube. The gas must be $CO_2$ as it is absorbed by KOH. This shows that $CO_2$ is released during fermentation. This is different from anaerobic respiration because here the respiratory substrate is outside the cell (glucose solution). Alcohol is formed during fermentation.

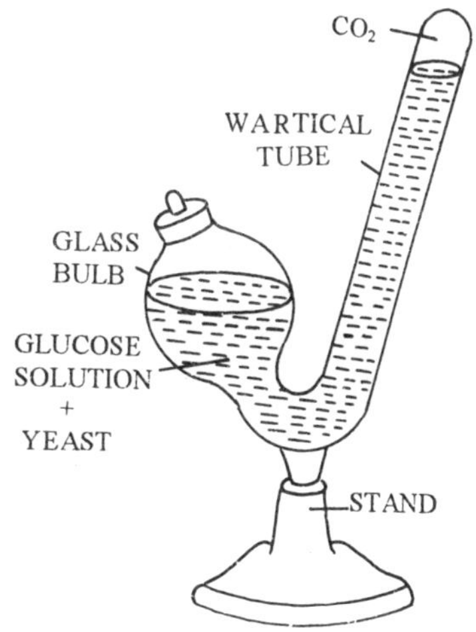


**Fig 14.33 Kuhne's fermentation flask**

### *54. Release of $Co_2$ during aerobic respiration (conical flask method)*

**Materials :** Conical flask, A twice bent tube, test tube, thread, germinating seeds, KOH solution, vesline beaker, saline water.

**Procedure :** Take a conical flask and place some germinating seeds in it. Close the mouth with a single holed cork. Introduce a twice bent tube through the hole. Take a small quantity of KOH solution in a test tube and hang it inside the flask with the help of a thread. The thread will be held tight with the help of the cork. Let the free end of the bent tube be dipped in a beaker containing saline water. Apply vaseline to the cork to make it air tight. Let the apparatus stand for some time

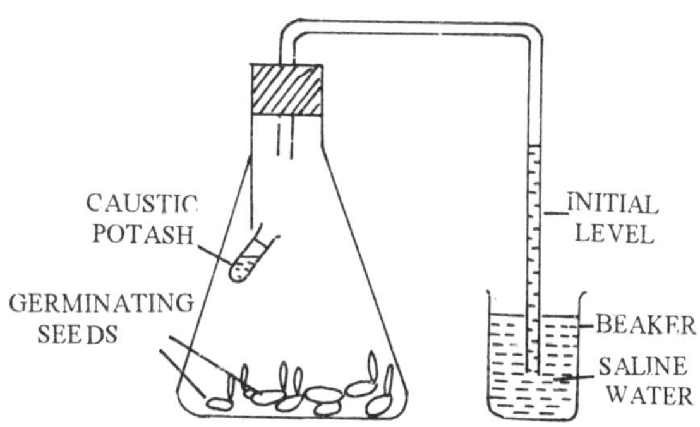


**Fig 14.34 Release of $CO_2$ during aerobic respiration**

**Observation :** Water level in the tube rises after some time.

**Conclusion :** As the germinating seeds respire to release $Co_2$ which is absorbed by KoH. This creates a kind of suction pressure which is transfered along the tube into the beaker containing saline water making it to rise.

### *55. Evolution of heat during respiration*

**Materials :** Two thermos flasks, germinating seeds, mercuric chloride, rubber cork etc.

**Procedure :** Take two thermos flasks. Keep germinating seeds in one and mercuric chloride treated (killed) seeds in the other. With the help of single holed corks introduce a thermometer into each of these flasks and close the mouth (of the termo flask). Note the inital temperature in both the flasks. After 45 minutes note the temperature again. Record your observations.

**Observation :** In the thermos flask with killed seeds there will be neither increase or decrease in termperature as recorded by the thermometer. In the other flask having germinating seeds there will be an increase in temperature as recorded by the thermometer.

**Conclusion :** During respiration there is release of energy, while most of it used for the synthesis of ATP, some amount escapes as heat and this results in increased temperature in the flask with germinating seeds. In the other flask where the seeds are killed, there is no increase in temp as there is no respiration.

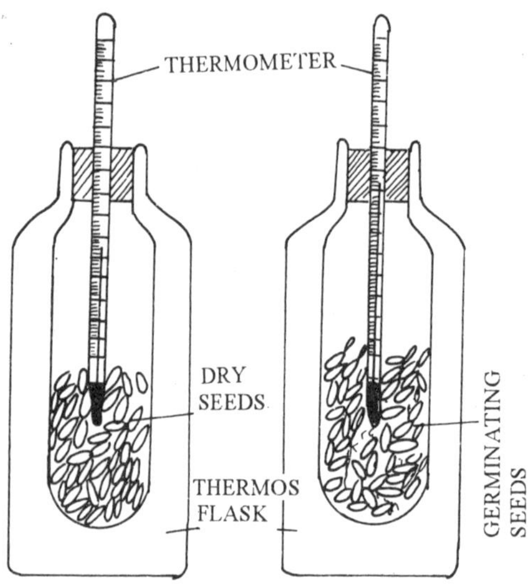


**Fig 14.35 Evolution of heat during respiration**

### *56. Demonstration of use of $O_2$ during Respiration*

**Materials :** Conical flask, beaker, bent glass tube, trough, mercury, germinating seeds, caustic potash stand etc.

**Procedure :** Keep some germinating seeds in a conical flask. Also keep a small vial containing KOH inside the flask. Now close the mouth of the conical flask with the help of a cork having a single hole. Through this hole introduce a twice bent glass tube. Let the the free end of the tube be inserted in a beaker containing mercuryh. Allow the experiment to stand for some time.

**Observation :** Mercury kept in the beaker rises in the glass tube up to 15 cm height and then stops.

**Conclusion :** Germinating seeds respire and release $Co_2$ which is immediately absorbed by KOH. This creates a suction force and mercury rises up in the glass tube. After reaching a height of 15 cms rise of level of mercury stops indicating the stoppage of respiration and consequently the production of $CO_2$.15 cms height of mercury in the tube is approximately one fifth of normal height at which atmospheric pressure maintains it. This means that the seeds must have utilized 1/5th of air. The share of $O_2$ in atmospheric gases is also 1/5th. It may therefore be assumed that seeds have absorbed $O_2$ (for respiration) which accounts for 1/5th of air.

The rise in level of mercury is equivalent to the amount of $CO_2$ produced. This means the production of $Co_2$ during respiration is about 1/5th of the volume of atmosphere. During respiration the amount of $O_2$ utilized is equivalent to the amount of $Co_2$ released ($C_6H_{12}O_6 + 6O_2 + 6H_2o$ $6Co_2 + 2H_2o$). Hence this indicates that 1/5th of volume of air i.e, the entire $O_2$ content is used up. As there is no further amount of $O_2$ left in the conical flask respiration stops and the rise in the level of mercury also stops.

### *57. Demonstration of evolution of $CO_2$ during aerobic respiration by gas flow method*

**Materials :** Potted healthy plant, bell jary black cloth, wide mouth bottles, 'U' tube, soda lime, bent tubes, lime water, baryta water, aspirator, vaseline etc.

**Procedure :** Take a bell jar and invert it over a healthy potted plant. Close the top of the bell jar with a two holed cork. Connect two bent tubes to these holes. To one tube connect a bottle containing baryta water. This bottle also should be fitted with a two holed cork. Insert a bent tube into this bottle with the other end of the tube being connected to an aspirator. Now the other bent tube from the bell jar should be connected serially to two bottles having lime water and then onto a 'U' tube containing soda lime. Smear vaseline to all the connections and cover the bell jar with a black cloth.

Set up the apparatus in such a way that air necessary for respiration first enters the 'U' tube followed by two bottles containing lime water and then into bell jar. As a result the air that enters the bell jar will be free of $CO_2$ as it gets absorbed by lime water in bottles. As the bell jar is covered by black cloth there will be no photosynthesis and hence no chance of $CO_2$ released during respiration to be utilized for photosynthesis. The air now enters the bottle containing baxyta water and then to the aspiration.

**Observation and conclusion :** Baryta water turns milky showing that the air from bell jar (with the plant) has $CO_2$. The air entering the bell jar is $Co_2$ free while the air leaving bell jar has $CO_2$. Hence the $CO_2$ must have been released by the plant during aerobic respiration.

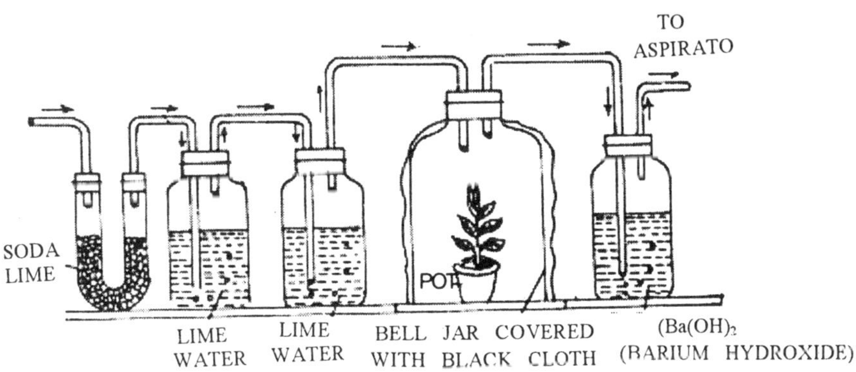


**Fig 14.36 Evolution of $CO_2$ during aerobic respiration (Gas flor method)**

### *58. Evolution of $CO_2$ during an aerobic respiration*

**Materials :** Same as experiment 57 and pyrogallol.

**Procedure :** Set up the experiment in the same manner as in 57, except instead of two bottles having lime water, replace lime water, in one bottle with pyrogallol. As the air enters thsese bottles, one (lime water) absorbs $CO_2$ and the other (pyrogallol) absorbs $O_2$ so that the air entering bell jar has neither $CO_2$ nor $O_2$. Still the air that leaves bell jar has $CO_2$ indicating respiration in the absence of $O_2$. This demonstrates anaerobic respiration.

***59. Measurement of Repiratory quotient (RQ) of different respiratory substances using Ganong's respirometer***

**Materials :** Ganong's respirometer. Water, saline water, stand, cotton, different germinating seeds such as wheat, rice (carbohydrate), castor, mustard (fat), pea (protein), succulent plant such as *Opuntia* or *Bryophyllum*.

**Procedure** : Ganong's respirometer basically consits of a vertical graduated tube connected a stopper. The neck of the bulb as well as the stopper have a small hole each so that they are in line at a particular level. The other end of the graduated tube is attached to a levelling tube with the help of rubber hole. Take some quantity of saline solution and pour it into the levelling tube so that both tubes get filled about one half their capacity. Now open the stopper and keep some wet cotton in the bulb together with some germinating seeds (any one mentioned above - in materials). Place the stopper and rotate it slightly in the mouth of the bulb so that the holes in them are in line. The air pressure inside the bulb now equals that of outside. Adjust the levelling tube (lowering or raising) in such a way that the level of saline water in the graduated tube reaches zero level Rotate the stopper now to cut off contact with external air. Let the apparatus stand for some time.

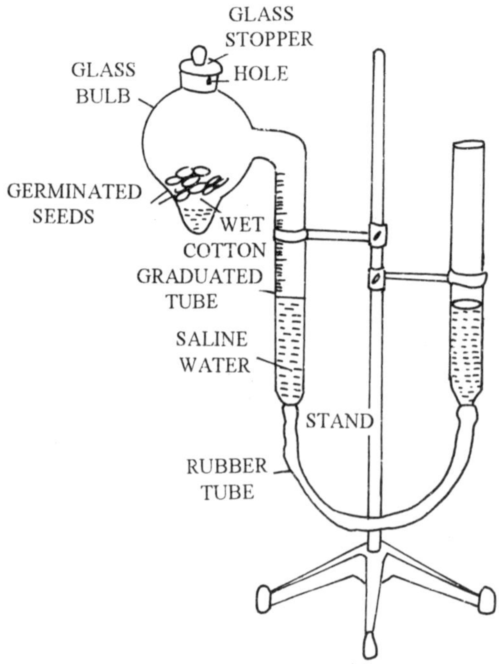


**Fig 14.37 Ganong's respirometer**

**Observation :** The Level of saline solution in the graduated tube remains stationary

**Conclusion :** No change in the level of saline solution indicates that the amount of $O_2$ used is equivalent to the amount of $CO_2$ released. There will be no change in volume of air in the bulb. Hence it does not press down nor suck the saline solution from the graduated tube. Thus the RQ is equivalent to 1 or unity. RQ is calcualted as follows.

$$RQ = \frac{\text{Amount of } Co_2 \text{ evolved}}{\text{Amount of } O_2 \text{ used}}$$

In case of **Carbohydrates** (Wheat, Rice,) such as sugars vol of $CO_2$ evolved is equivalent to $CO_2$ used.

$$C_6HO_6 + 6O_2 + 6H_2O \rightarrow 6CO_2 + 12H_2O$$

$$RQ = \frac{\text{Amount of } CO_2 \text{ evolved}}{\text{Amount of } O_2 \text{ used}} = \frac{6}{6} = 1 \text{ or unity}$$

When **fats** are respiratory substrates (castor, mustard etc) the level of saline water in the graduated tube will rise. Fats have less $O_2$ and hence require more $O_2$ for oxidation. $O_2$ used in respiration is more than $CO_2$ evolved and RQ will be less than unity. When palmition is the respiratory subbstrate -

$$2C_{51}H_{98}O_6 + 145O_2 \longrightarrow 102CO_2 + 98H_2O$$

$$RQ = \frac{102}{145} = 0.7$$

When the respiratory substrate is a protein (pea, gram etc) the level of saline solution will rise because proteins also require more $O_2$ in respiration than the amount of $CO_2$ evolved. As a result volume of air in the bulb gets decreased and the saline water rises being sucked by the air. (Usually RQ for proteins will be 0.8 - 0.9)

In case of **organic acids** being respiratory substrates, more amount of Cos will be released than O2 utilized. As the volume of air increases in the bulb it presses down the saline solution and its level will go down. Here RQ will be more than unity.

$$C_4H_6O_5 + 3O_2 \longrightarrow 4CO_2 + 3H_2O$$

$$RQ = 4/3 = 1.33$$

**Caluclation of actual RQ in an experiment using Ganong's respirometer**

1. Note the initial level of saline solution in the graduaated tube ( $l_1$)
2. Final level ($l_2$)
3. The difference between the two $l_1 - l_2 = V_1$ (The value of $V_1$ may be negative or positive)

Insert a piece of KOH into the graduated tube. As the total $CO_2$ is absorbed the level will rise further. This rise may be called $l_3$.

The difference between $l_2$ and $l_3$ $(l_3 - l_2) = V_2$

$V_2$ denotes the actual volume of $CO_2$ evovled during respiration. While $V_2 + V_1$ gives the volume of $CO_2$ used during respiration.

### *60. Comparison of Respiration and Photosynthesis*

**Materials :** Two long necked flasks, petridishes, stands, mercury, KOH pellets, green leaves, petals etc.

**Procedure :** Take two flasks A and B. Place few fresh moist green leaves in flask 'A' and moist petals in 'B'. Invert the flasks over troughs of mercury so that the neck is immersed in mercury. Fix the flasks to separate stands with the help of clamps. Keep the apparatus under diffuse light for about 10 - 12 hours. At the end of this period introduce KOH pellets into both the flasks and observe.

**Observation :** There is very little rise in the level of mercury in flask 'A' while in flask 'B' there is a considerable rise. In flask 'A' green leaves photosynthesize and utilize $CO_2$ rleased during respiration. Hence there is little if any free $CO_2$ left in the flask. In flask 'B' where petals are kept the entire $CO_2$ is in the flask when KoH is introduced it obsorbs $CO_2$ and hence there is rise in the level of mercury.

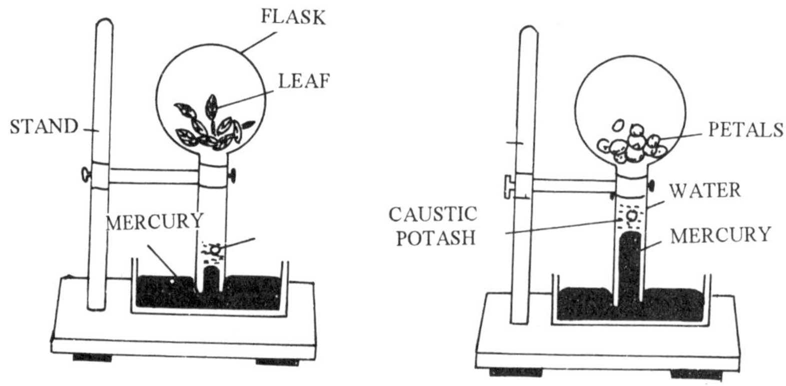


**Fig 14.38 Comparison of Respiration and Photosynthesis**

## ENZYMES

### *61. Test of Oxidase*

**Materials :** Fresh apple, tray,

**Procedure :** Cut thin slices of apple and place them on a tray. The pieces turn brown after some time. This shows that the phenolic compounds get oxidised in the presence of atmospheric oxygen due to the action of enzyme oxidase.

The enzyme oxidase catalyses the oxidation of the substrate by removing hydrogen from them and transfering it to molecular oxygen as follows.

$$AH_2 + O \quad A + H_2o$$

### *62. Test for Peroxidase*

**Materials :** Potato tuber, Buchner's funnel, beaker, mortar, Buchner's funnel, muslin cloth, test tubes, 1% guaiacum (Benzidine) solution, $3H_2o_2$.

**Procedure :** Crush a fresh potato in a mortar. Filter the slurry through two layers of muslin cloth in a Buchner's funnel. Tkae the filterate (3 ml) in a test tube and add 0.5 ml of 1% Benzidine (dissolve 2 gms of guaiacum in 100 ml of absolute alcohol). Immediately add to the test tube a few drops of 3% $H_2O_2$ solution. Note down the change in colour of the mixture.

**Observation :** The colour rapidly changes to blue

**Conclusion :** Phenolic compounds get oxidized rapidly in the presence of $H_2O_2$. This shows the presence of peroxidase in the potato tubers. If boiled potato (the enzyme will be denatured) is taken blue colour will not appear. The reaction is as follows :

$$AH_2 + H_2o_2 \quad A + 2H_2o$$

*63. Test for amylase*

**Materials :** Fresh potato, starch solution, germinating barley grains, citrate - phosphate buffer (pH 5.6), Iodine solution, test tube, spirit lamp, water etc.

**Procedure :** This involves two steps (i) preparation of enzyme extract and (ii) assay for enzyme activity.

**Preparation of enzyme extract :** Mash germinating barlye grains in cold citrate phosphate buffer (pH 5.6) and filter the extract through two layers of chees cloth. The filterate serves as the enzyme source. Keep it in ice bath.

**Enzyme assay :** Take two clean test tubes A and B. Add 2ml of enzyme extract to each of the test tubes. Boil the enzyme extract in A in a water bath. Cool it to room temperature. Now to bath tubes add freshly prepared potato starch solution (1 ml) and shake gently. Keep the tubes aside for 10 minutes. Test the contents of A and B separately at periodic intervals with Iodine solution for the presence of starch.

**Observation and Conclusion :** In tube A the contents always turn blue indicating the presence of starch. In tube 'B' the blue colour will be intense in the beginning but later it turns lighter and finally no colour formation will be there. This indicates the gradual digestion of starch due to the activity of amylase. In tube 'A' the enzyme will not be active as it gets denatured due to boiling.

## NITROGEN METABOLISM

*64. Demonstration of nitrates in green plants :*

**Materials :** A healthy potted plant, slide, cover slip calcium nitrate solution (1.5%), diphenylamine (1%) , microscope.

**Procedure :** Select a healthy potted plant. Add to the pot 100 ml of 1.5 % calcium nitrate solution. Keep the plant in darkness for 72 hours and then transfer it to bright light. Cut transverse section of lamina and petiole of the leaf and mount them in few drops of diphenylamine sulphate solution (1% diphenylamine dissolved in 50% $H_2SO_4$). Cover them with cover glass and observe under the microscope.

**Observation :** A deep blue colour appears

**Conclusion :** The colour indicates the presence of nitrates.

*65. Experiment to demonstrate that atmospheric $N_2$ is not directly absorbed by green plants :*

**Materials :** Glass troughts (one big and one small), bell jar, dilute sulpuric acid, petridish, funnel tube, sterilized $N_2$ free sand distilled water.

**Description of the apparatus :** The apparatus consists of a large glass trough (A) which is half filled with dilute sulphuric acid. A bell jar (B) is inverted inside the trough and placed on three corks so that it is raised above the floor of the trough. A small trough (C) is placed inside the bell jar on a large petridish so that it is not in contact with $H_2SO_4$ in the large trough. Trough 'C' is filled with water and a small glass pot is kept inside the trough (C). The inverted bell jar has a small inlet tube in the side which is connected on the inside to trough 'C' and on the outside to a funnel. This facilitates addition of water from time to time. The top of the bell jar is closed with a cork having an inlet tube which allows $CO_2$ inside the bell jar.

**Procedure :** Collect some legume seeds and estimate their total $N_2$ content by chemical analysis collect the ash also and estimate the $N_2$ content. Take some quantity of sterilized $N_2$ free sand and add it to the small glass pot inside trough 'C' together with some quantity of ash. Sow some legume seeds (the seeds from the same plant on which chemical analysis has been done) in the pot and allow them to germinate. Let them grow into small plants. After a definite period of time (15-12 days ) pull out the plant along with roots. Estimate the total $N_2$ present in the whole plant. Compare it with the amount of $N_2$ already estimated in the seeds.

**Observation and Conclusion :** Comparison of the $N_2$ content shows that the plant does not have more $N_2$ than the seeds. This shows that the plant has not absorbed gaseous $N_2$ from atmosphere. Higher plants can only absorb $N_2$ in the form of nitrates or nitrates. In this experiment the sand from which roots absorb nutrition is $N_2$ free. The only available form of $N_2$ for plants is gaseous $N_2$. Dilute $H_2SO_4$ absorbs traces of ammonia if any in the bell jar.

## GROWTH AND HORMONES

### *66. Measurement of Growth by arc auxanometer*

**Materials :** Arc auxanometer, weight, thread, a potted plant.

**The apparatus :** The arc auxanometer consists of a tripod stand and a pulley connected to a pointer which moves on an arc scale. A thread is made to pass over the pulley tightly. One end of the pulley is tied to a weight.

**Procedure :** Take a healthy potted plant anJ tie its tip to thread which is held over the pulley in such a way that the thread does not slacken. Keep it aside and observe.

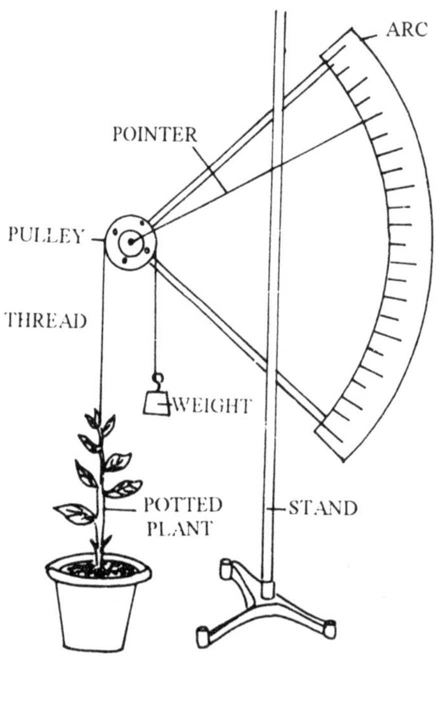


**Fig 14.39 Arc Auxamometer**

**Observation and Conclusion :** As the stemp top grows, it will push the thread up and the tight thread will turn the pulley. The pulley in turn will move the pointer over the graduated arc scale. The extent of movement from the initial to the final in a given period of time can be measured. The distance moved by the pointer does not give the actual linear growth, because the arc is divided into 90° (one quarter of a circle i.e., 360 / 0.25) and not linear measurements. Further the measurement on the graduated arc is magnified because of the length of the pointer. The actual growth is calculated as follows-

Inital position of pointer = A

Final position of pointer = B

Actual distance travelled by the pointer during a given period of time = B - A

Acutal growth = B - A x Radius of pulley

in length length of pointer (from the centre of the pulley

For example if the radius of pulley is 10 cms and the length of the pointer is 50 cms the magnification is 10 times. If the pointer has moved 5 cms in 20 hours the actual growth is as follows.

5 x 10 = 1 cm in 20 hours

50 Therefore Growth per hour = 0.5 mm

### 67. *Measurement of growth by Pfeffer's Auxanometer :*

**Materials :** Pfeffer's auxanometer, potted plant, thread etc.

**The apparatus :** Pfeffer's auxanometer consists of a vertical stand and a self rotating drum over which a smoked paper is glued. The drum rotates on its axis through a clock work mechanism at the rate of one turn per every hour. Attached to the stand on a shaft are two pulleys one small and the other bitg. The smaller pulley is attached to the central axis of the larger pulley. A thread passes over the small pulley, one end of which is tied to a weight. Another thread also passes over the large pulley with weights tied to both ends. A pointer is attached to one end of this thread slightly above the weight. This pointer touches the smoked paper wrapped around the rotating drum.

**Procedure :** Take a healthy potted plant and tie its tip to the free end of the thread coming over the small pulley. Observe.

**Observation and Conclusion :** As the stem tip grows the small pulley moves together with large pulley. This results in the pointer continuously marking a spiral line on the rotating drum's smoked paper.

The perpendicular distance between two adjoining spirals is the hourly growth of the plant. Actual growth rate may be calculated taking into consideration the radii of both the pulleys and the rate of rotation of drum.

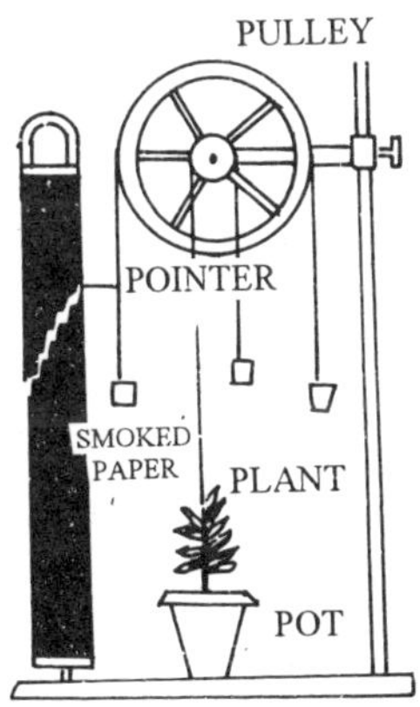


**Fig 14.40 Pfeffer's Auxanometer**

### 68. *Measurement of growth by space markers*

**Materials :** Indeliable ink (India ink), germinating seeds

**Procedure :** Select few germinating seeds where radicle has emerged out into the tap root. Using a thin nib mark graduations at regular intervals a little behind the root tip. Observe.

**Observation and Conclusion :** It is observed that the distance between the graduations have increased. This increased distance is the actual linear growth of the root.

### 69. *Effect of Auxins (Indole acetic acid - IAA) on plant growth*

**Materials :** Oat grains, Auxin solution (10 mg/l of IAA), 10% sucrose solution, water, circular filter paper, graph sheet etc.

**Procedure :** Soak some oat (*Avena sativa*) grains in water and place them on a wet filter paper in a petridish in dark. After about 2 days when coleoptiles are 2 - 3 cms long, remove their tips and discard them. Cut sections of about 5 mm long from the decapitated coleuptile tips. Collect about 15 sections, and keep them in three separate petridishes (each having five sections) as follows :

Petridish A Water

Petridish B Sucrose solution

Petridish C Sucrose solution + 1 ml auxin solution.

Keep the petridishes back in darkness and observe.

**Observation :** It is observed that *Avena* coleoptile pieces kept in petridish C have grown much longer than those in A and B. The shortest length is found in A.

**Conclusion :** Coleoptile elongation is minimum in 'A' (water) as it has no nutrition. While slight growth is found in B because of sucrose solution. In 'C' however these is maximum elongation as auxins promote coleoptile growth.

**Note :** Different concentrations of IAA may be tried to find out their effect on growth

### *70. Effect of Gibberellins on plant Growth*

**Materials :** Seeds of pea, tomato, lettuce etc, petridish, circular filter paper, gibberellic acid solution (100 mg/litre), measuring cylinder etc.

**Procedure :** Germinate seeds in dark on wet filter paper placed in a petridish. When the seeds have germinated for about 48 hours, transfer them to petridishes - (a) filter paper with water and (b) filter paper with gibberellic acid solution. Keep the petridishes in light for 72 hours and observe.

**Observation :** It is seen that hypocotyls in the petridish with gibberellic acid solution become much more elongated than the one with only water.

**Conclusion :** Gibberellin has an effect of elongation of cell and hence the hypocotyls become longer in length.

### *71. Effect of Cytokinin on expansion of Cotyledons*

**Materials :** Seeds of cucumber, pumpkin or ash gourd, petridish, filter paper, 10 mg kinetin solution, water, graph paper etc.

**Procedure :** Germinate the seeds on wet filter paper placed on a petridish in dark. Select germlings with about 2 - 3 mm of hypocotyl and remove the seed coat. Separate the cotyledons and float them on petridishes - (a) containing only water and (b) containing kinetin solution. The abaxial surface of the cotyledons should be in contact with kinetin solution. Keep them aside for 72 hours and observe.

**Observation :** It is seen that cotyledons kept in kinetin solution have expanded much more than those kept in water. This can be tested by placing them on graph paper.

**Conclusion :** Kinetin a growth hormone is known to induce rapid cell division thus bringing about expansion of the cotyledons.

## PLANT MOVEMENTS

### *72. Geotropism (Horizontal Klinostat)*

**Materials :** Klinostat, a healthy plant

**The appratus :** Horizontal klinostat essentially consists of a clockwork mechanism. Instead of the hands at the region of the dial there is an axis to which is attached a small metallic tumbler. The clock is fitted to a stand. When the clock is working the tumbler slowly rotates. It remains stationary when the clock is not working.

**Procedure :** A small potted plant is kept in the tumbler as shown in the figure. Observe what happens to the plant under the following conditions

(a) when the tumbler is rotating (clock working)
(b) When the thumbler is not rotating (clock not working)

**Observation :** When the tumbler is stationary i.e., clock not working, after few days the shoot bends up while the root bends down (downward curvature of the root may be observed by removing the soil). When the tumbler is rotating, the plant continues to grow horizontally i.e, parallel to ground. There is no upward curvature in the stem and downward curvature in the root.

**Conclusion :** When the clock is not working auxins accumulate in the lower surface (ground facing side) of stem and root. Auxin accumulation has opposite effects on stem and root. In stem auxin accumulation increases cell division while in root it decreases cell division. As a result of this, shoot bends up because there is more growth on the lower surface than the surface away from ground. In root the auxin accumulated lower surface has less growth than the surface away from gravity hence it bends down. The unequal accumulation is due to gravity.

When the tumbler is rotating there will not be gravitational effecti on any part of the plant as auxins can not accumulate in any one surface. As the plant is also rotating there is equal distribution of auxins. Hence the plant (shoot or root) does not bend but grows horizontally.

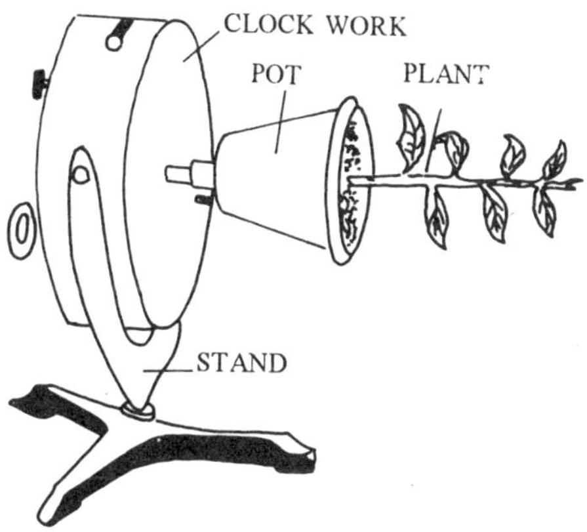


**Fig 14.41 Horizontae Klinostat**

*73. Effect of Geotropism (Vertical Klinostat)*

**Materials :** Vertical klinostat, germinated seeds with hypocotyl emerging, cork, pin, etc.

**The apparatus :** A vertical klinostat is essentially like a horizontal klinostat except that the face of the clock work mechanism will be directed up. In addition, a disk is attached to the rotating axis instead of tumbler. When the clock is operating the disc rotates. Corks are fitted at regular intervals near the periphery of the disc.

**Procedure :** Attach a few germinated seeds to the cork with the help of pins. Observe the behaviour of seedlings under the following conditions :-

(a) When the disc is rotating slowly
(b) When the disc is not rotating
(c) When the disc is rotating fast

**Observation :** When the disc is rotating slowly i.e 2.5 turns per hour the roots will diverge and shoots converge at an angle of 45°. The bending is due to the fact that centrifugal force (rotating force) and gravitational force act together and bring about the slanting growth.

When the disc is not rotating only geotropic force will act on the plant. Shoot will grow up and root down.

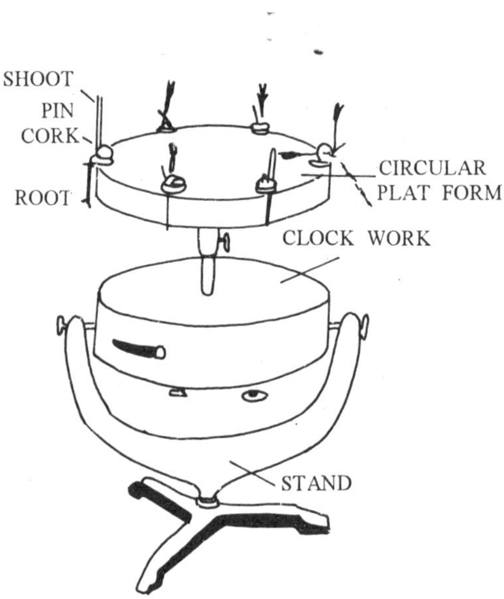


Fig 14.42 Vertical Klinostat

When the disc is rotating fast the geotropic force is nullified by the centrifugal force. As result the plant grows horizontally the root away from the disc and the shoot towards the centre of the disc.

### *74. Phototropism (Heliotropism)*

**Mateials :** Phototrophic chamber (box), A potted plant.

A phototrophic chamber is a small wooden box painted black inside. It has a hinged door at the top (to open and close) and a small window on the side.

**Procedure :** Keep a potted plant in the phototrophic chamber close the lid. Keep the box in such a way that the window of the box is directed towards the source of light. Let the box stay in place for about 2 - 3 days.

**Observation :** It is observed that the shoot bends towards the window of the box, comes out and then once again bends up and grows vertically.

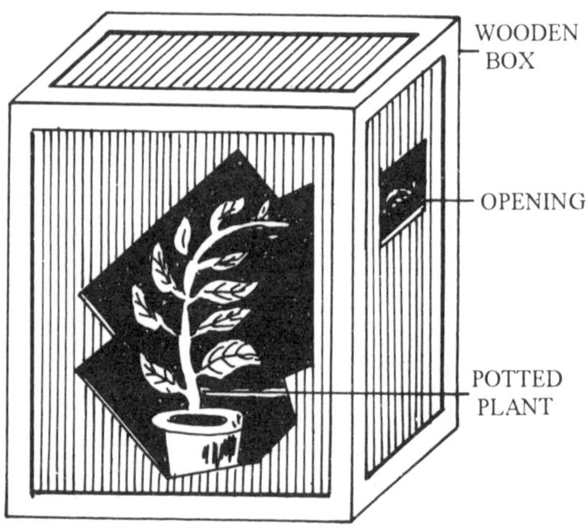


Fig 14.43 Phototropic Chamber

**Conclusion :** The plant gets light only through the side window. Unidirectional light falling on the side of the stem destroys the auxins on that part of the stem. Auxins accumulate in the shaded region. As a result there is more growth in the shaded region than the lighted region and the shoot bends and comes out of the window of the box. When the shoot comes out of the box the light strikes the upper surface of the horizontal stem making auxins to concentrate on the lower surface. Hence the stem bends up then onwards as the light is not undirectional on the apex it grows up. This is known as phototropism i.e., growth towards light. While shoots are positively phototropic roots are negatively phototropic i.e. they grow away from light (towards gravity)

# GERMINATION

*75. Viability (Germinability) test for seeds :*

**Materials :** Barley seeds, water, petridish, filter paper, 0.1 solution of tetrazolium chloride.

**Procedure :** Soak barley seeds in water for about 2-3 hours. Cut the seed exactly into 2 parts slicing through the embryo. Keep the halves in petridishes containing a solution of tetrazolium chloride. The seeds should be completely immersed in the solution. Keep the petridishes in dark for about 45 minutes and observe.

**Observation :** Some of the seeds have stained embryos while others are colourless.

**Conclusion :** Viable seeds respire and can change the colourless tetrazolium compound to a coloured (reduced) compound formazon. Formazon is red in colour and thus the embryos are stained. In non viable seeds there is no respiration and hence the tetrazolium salt remains colourless. Percentage of germination of seeds can be calculated as follows :-

Number of seeds used for experiment = A
Number of seeds stained = B
Percentage germination B/A x 100
Microchemical Tests

# CARBOHYDRATES

*76. Test for Cellulose*

**Materials :** Paper or Cotton fibres, Iodine solution, water 75% sulphuric acid, slides, cover slips, microscope etc.

**Procedure :** Place a few cotton fibres or very thin shreds of paper in a drop of water. Observe the cell wall under the microscope. Add a few drops of Iodine and a few drops of $H_2SO_4$.

**Observation :** The fibres turn yellowish brown first and then turn blue.

**Conclusion :** Blue colour indicates the presence of cellulose

*77. Test for Cutin*

**Materials :** *Ficus* or *Nerium* leaf, microscope, slides, coverslips, watch glass, water, KOH solution etc.

**Procedure :** Cut cross sections of *Nerium* leaf, mount the sections on a slide, stain them with KOH solution. Put coverglass and observe the colour of the outermost deposition on the epidermis.

**Observation :** The deposition takes a yellow colour.

**Conclusion :** Yellow colour indicates the presence of cutin.

*78. Test for Hemicellulose*

**Materials :** Soyabean seeds, Iodine, water, slide, cover glass, glycerine etc.

**Procedure :** Cut a thin section of the seed, place it in water and stain it with iodine solution for few minutes. Observe under the microscope.

**Observation :** Sections particularly the walls take a blue colouration.

**Conclusion :** Blue colouration indicates the presence of hemicellulose.

*79. Test for Suberin*

**Materials :** Natural cork, sudan IV (alcoholic), 50% ethyl alcohol, slide, cover glass, water etc.

**Procedure :** Cut thin sections of cork. Stain them with sudan IV for about 15 minutes. Wash excess stain with 50% ethyl alcohol. Transfer the sections to water for about 5 minutes and finally mount in glycerine. Put a cover glass over the section and observe under the microscope.

**Observation and conclusion :** All the suberized portions take red stain, indicating the presence of suberin in the cell wall.

*80. Test for Lignin*

**Materials :** Wood shavings, sections of woody material, 1% alcoholic phloroglucin, 25% Hcl, 1% neutral aqueous $KmNO_4$, Amonium hydroxide or sodium bicarbonate, slide, cover glass, water etc.

**Procedure :** (1) Treat the material (section or wood shavings), with phloroglucin on a slide and put a coverglass. Flood the section with HCL along the edges of the cover glass. Observe.

(2) Treat the material with potassium permanaganate ($KmNO_4$) solution for about 15-20 minutes. Wash the material with 25% Hcl. Follow it up with several washes of water. Add a few drops of either $NH_4OH$ or sodium bicarbonate solution.

**Observation and Conclusion :** In the first method lignified walls take up a red-violet colour. In the second method a deep red colour develops.

*81. Test for Mucilage :*

**Materials :** Seed coat of linseed, 10% $CuSO_4$ solution, 10% KOH solution, water, slide, cover glass etc.

**Procedure :** Cut thin sections of the seed coat and soak them in 10% $CuSO_4$ solution for about 20 minutes. Wash them in water and transfer to 10% KOH solution. Observe the sections under microscope after mounting them in glycerine.

**Observation and conclusion :** Cells having mucilage are stained bright blue.

*82. Test for Latex*

**Materials :** Latex of *Calotropis*, *Asilepias*, *Euphobia* etc sucrose, Conc $H_2SO_4$, test tube, water etc.

**Procedure :** Extract the latex in alcohol (mix alcohol and latex thoroughly; centrifuge and take the supernatant) and add an equal amount of sucrose and sulphuric acid.

**Observation and Conclusion :** The colour turns pinkish purple. This indicates the presence of latex.

*83. Test for Glucose (Fehling's test)*

**Materials :** Glucose solution, Test tube, spirit lamp, freshly prepared solution A and B.

**Procedure :** Mix about 1.5 ml each of Fehyling's solution A and B in a test tube and add 3ml of glucose solution. Boil on spirit lamp. Cool it and observe.

**Observation and Conclusion :** A brick red pricipitate develops indicating the presence of glucose. Fehlings solution has cupric sulphate, sodium hydroxide, and potassium sodium tartarate. When mixed with glucose and boiled, the aldehyde groups of glucose are oxidized while the cupric salt gets reduced to cuprous oxide which forms a brick red pricipitate insoluble in water.

$$C_6H_{12}O_6 + 2Cu\,(oH)_2\ C_6H_{12}O_2 + Cu_2o + H_2o$$

*84. Test for Glucose (Bendecit's test)*

**Materials :** Glucose solution, Bendict's solution, spirit lamp, test tube etc.

**Procedure :** Add 1 ml of glucose solution to 3 ml of Benedict's solution taken in a test tube; warm the mixture on the flame of spirit lamp untill a red brown colour appears.

**Observation and conclusion :** The colour changes from blue to green and finally to rust brown indicating the presence of reducing sugars.

*85. Test for Sucrose :*

**Materials :** Sucrose solution, Fehling's solution A and B, NaOH solution, HCl litmus paper, spirit lamp

**Proceudre :** Take 3 ml of sucrose solution in a test tube and few drops of conc. HCl and boil on the flame of spirit lamp, cool. Add a little NaOH to neutrialize the mixture. Test the pH with litmus paper. Mix Fehlings solution A and B. Add 3 ml of this mixture to the neutralized mixture of sucrose solution. Boil again till a brick red precipitate appears.

**Observation and Conclusion :** Sucrose is a non reducing sugar, but on acid hydrolysis it gets broken up into glucose and fructose which are reducing sugars. These on treatement with Fehling's solution give a brick red precipitate.

*86. Test for Starch*

**Materials :** Starch, supspension, test tube, Iodine solution, spirit lamp.

**Procedure :** Take 5 ml of starch suspension in a test tube and add equal volume of idoine solution. watch the change in colour.

**Observation and Conclusion :** Starch gives a blue colour on reacting with iodine. The colour disappears on boiling but reappears on cooling. Starch molecules form a complex with iodine solution under normal temperature.

## PROTEINS

*87. Biuret Test*

**Materials :** Gram flour, test tube, 40% NaOH, 1% CuSo4 solution.

**Procedure :** Mix some quantity of gram flour and prepare a solution. Take 5 ml of this solution in a test tube and add 2 ml of NaOH and one drop of $Cuso_4$ solution. Mix the contents by stirring and note the colour change.

**Observation and Conclusion :** A pale purple or pink colour appears indicating the presence of proteins. Biuret reagent reacts with the peptide bond of proteins and produces the colour.

*88. Xanthoproteic Test*

**Materials :** Gram flour, con. $HNO_3$, conc NaOH solution, spirit lamp etc.

**Procedure :** Mix a little gramflour with water and prepare a solution. Take about 3 ml of this solution in a test tube and add equal volume of conc. $HNO_2$. A thick white precipitate appears. Boil the solution, the precipitate turns canary yellow. Cool the test tube with cold water and add NaOH to neutralize and then to make it alkaline. Observe the colour.

**Observation and Conclusion :** Yellow colour turns to orange showing the presence of proteins. Aromatic aminoacids present in the protein when heated with $HNO_3$ form nitroderivative of aromatic aminoacids. These nitroderivatives on treatment with NaOH form salts which are orange coloured.

## FATS

### *89. Test for Fat*

**Materials :** Seeds of almond, sunflower etc., water, osmic acid, sudan III (0.5% in 70% alcohol), glycerine etc.

**Procedure :** Cut thin sections of the material; stain them with sudan III for about 10 minutes. Wash the sections with water. Observe the colour under microscope.

**Observation and Conclusion :** Materials turn red on microsccpic observation. It is seen that the oil droplets have taken the red stain. In another method a few drops of osmic acid is added to the material taken in a test tube. Formation of black colour indicates the presence of fat.

### *90. Test for Fat (oil)*

**Materials :** Olive oil, water, petroleum ether, sudan III (0.5% dye in 70% alcohol)

**Procedure :** Take a few ml of the oil in a test tube and add a few drops of sudan III. Shake well. The oil turns red. Pour half the contents into another tube. Add water to one tube and petroleum ether to another. Shake both the tubes gently and observe.

**Observation and Conclusion :** The red coloured oil does not mix with water hence it floats forming a red coloured ring over water. However when petroleum ether is added, oil dissolves in it and a homogenous red solution is formed.

### *91. Test for presence of fatty acids by saponification :*

**Materials :** Butter, 10% alcoholic KOH, test tube, litmus paper, spirit lamp, water, NaOH solution.

**Procedure :** Melt the butter and take 1 ml of the oil in a test tube and add 2 ml of alcoholic KOH. Boil the mixture for about 2 minutes and add 5 ml of distilled water. Boil again for 5-6 minutes. Cool and add a few drops of conc HCl to make the mixture slighly acidic. Test the pH with litmus paper. A layer of fatty acids is formed at the top. Transfer it to another test tube and add a dilute solution of NaOH and heat.

**Observation :** The solution turns soapy

**Conclusion :** The soapy solution consists of fatty acids. By the addition of NaCl solution, the soapy solution precipitates indicating the presence of the fatty acids.

### *92. Test to Determine the saponification value of Fat*

**Materials :** Butter, petroleum ether, 0.5 N alcoholic KOH, 0.5 N HCl, buretle, relex condenser, conical flask etc.

**Procedure :** Take one gram of butter in a conical flask (250 ml). Add 3 ml of fat solvent (petroleum ether) and shake to dissolve butter. Add 25 ml of KOH solution and attach a reflex condeser to the flask and heat the contents in a water bath for 30 minutes. Cool the mixture to the room temperature. Titrate the mixture using a burette against 0.5 NHCl. Note the reading; conduct a blank experiment (without butter) and note the reading.Find out the difference in values.

**Observation and conclusion :** The difference in titrated value between blank (without fat) and experiment (with fat) will be the number of mg of 0.5 N KOH required for the saponification of 1 gm of butter. This value is called the saponfication value for one gm of fat.

## NUCLEIC ACIDS

(Microchemical test)

### *93. Test for RNA and DNA :*

**Materials :** Onion roots, azure B solution, butanol (0.25 mg/ml in citrate buffer at pH 4), watch glass, microscope.

**Procedure :** With the help of a razor blade cut thin long sections of root tips of onion. Transfer them to a watch glass and stain them using Azure B solution for about 2 hrs. Wash the sections in distilled water and transfer them to another watch glass. Flood the sections with butanol for about 30 minutes. Mount the sections on a slide, put cover glass and observe under microscope.

**Observation and Conclusion :** Certain portions of the cells turn greenish blue and certain others turn deep blue. Greenish blue colour shows the presence of DNA and deep blue colour indicates RNA.

# PART FIVE
# CYTOLOGY

## CHAPTER 15

# CYTOLOGY

### TECHNIQUES OF CYTOLOGICAL STUDY

Cytological study in practicals involves the study of chromosomes which are visible during cell division. Both mitosis and meiosis are studied with the help of microscopic observation of cells during division. Young flower buds are chosen for meiotic study while freshly emerging root tips are used for mitotic study.

In flower buds anthers are selected as they have microspore mother cells which undergo reduction division (meiosis) to produce pollen grains ultimately. Ovule also can be chosen as a material for meiotic study as the megaspore mother cell (MMC) will be undergoing reduction division to produce megaspores. But it is not a suitable material as in the entire ovule there will be only a single MMC undergoing meiosis. In the case of anther however there will be a large number of microspore mother cells dividing meiotically.

Root tips are selected for the purpose of mitotic study as the meristematic cells there will be dividing mitotically. In some cases shoot tips (apical buds) also serve as an ideal material for mitotic study.

There are two methods for the microscopic study of flower buds and root tips. These are the *Smear* and the *Squash techniques.*

**Smear :** This involves the spreading of a single layer of cells for the purpose of study. Cells are selected when they are actively dividing. One of the prerequisite for such a study would be fixing the material at suitable stage. This is done by fixing material at different periods of time and then selecting one which is most suitable from the point of view of division stages. Microsporocytes of *Tradescantia, Rheo, Chlorophytum* etc and root tips of onion *Ficus* etc are killed (fixed) in carnoy's or other suitable fluid at an appropriate time.

Perfectly clean slides are very essential for smears. For this purpose they are immersed for a long time in a mixture of sulphuric acid and potassium bichromate. They are then throughly washed in running water and dried in a hot air oven or wiped clean with lint free cloth. Fresh young anthers are dissected out of the flower buds and placed in the centre of the slide. Using a sharp razor blade one end of the anther is cut and by gently applying the pressure microsporocytes are sqeezed out. The debris like anther wall and other cellular material is removed.

**Staining :** Few drops of acetocarmine stain is added on the microsporocytes and a cover glass is put on the material. A pressure is applied gently on the cover glass in order to spread the smear of the microsporocytes.

The slide is gently warmed over the spirit lamp. Excess stain may be removed with the help of blotting paper. The edges of the cover slip is sealed with melted wax in order to prevent evaporation. The slide is now ready for the observation of meiotic stages.

**Squash Technique :** This method is adopted for root top mitotic studies. An onion bulb is placed on a coupling jar filled with water. The lower stock portion of the bulb should be immersed in water. This willl facililate the quick emergence of roots. After the roots are about 1 cm long the tips should be cut.

**Pretreatment :** Before squashing, the root tips are to be pretreated with either (a) calchicine or (b) .8 hydroxy quinoline

(a) Prepare a 0.5% or 1% aqueous solution of commercially available colchicine. Take about 10ml (depending on the quantity of the material to be squashed) or more of colchicine solution in a watch glass and drop the root tips into it. Let the root tips remain in this solution for about 30 minutes at room temperature.

(b) Prepare an aqueous 0.29 solution of 8 hydroxy quinoline and treat the rot tips in this solution for 2½ to 3 hours at 12-14°C temperature.

**Staining :** After pretreatment roots should be washed in distilled water and treated with acetic alcohol (1:3) solution for 15-30 minutes. Acetic alcohol is a fixative and the roots may be preserved in this fluid for a long time.

Take few root tips in a watch glass and add 2% aceto orcein in HCL in the ratio of 9 drops of stain +1 drop of HCL for every two root tips. At this stage the watch glass should be warmed over the flame of spirit lamp until the fumes of acetic acid escape. Care should be taken not to boil the mixture. Cool the material for about 15-20 minutes. Now take a root tip on a clean side, place a drop or two of acetoorcein stain put a cover glass and gently press with the back of a pencil or any flat object. Reamove excess stain by blotting paper. Seal the edges by metled wax to prevent evaporation. The slide is now ready for observation.

**Permanent Mounting :** The methods mentioned above are for temporary microscopic observation. They are to be made permanent as follows if they have to be stored for a long time. Slides that are to be made permanent must at least be 24 hours old after squashing. Select such slides and pass them through terrtiary butyl alcohol and acetic acid series taken in a coupling jar as follows :

(a) 1:1 acetic acid : tertiary butyl alcohol
(b) 1:3 acetic acid : tertiary butyl alcohol
(c) only tertiary butyl alcohol.

Scrape and remove all the wax used for sealing the cover glass. Then keep the slides in (a) for 5-10 minutes. Within this period the cover glass slips off the slide. Do not discard the cover glass. Material will be adhering to both the slide and coverglass. Hence both of them should be processed. Transfer both the slide and coverglass separately to (b). In order to facilitate handling the delicate cover glas a tooth pick may be split a little at the tip and the cover glass edge inserted on to it. Leave both the slide and cover glass in (b) for 10 minutes and then pass them on to (c) for 5 minutes. The slide and cover glass are now ready for mounting.

**Mounting :** As both the slide and cover glass has the material, both of them should be used in mounting. But the cover glass should not be put on the same slide. As both these have the material there will be over lapping and the stages will not be clear. Hence both the slide and coverglass having the material should be mounted separately with another cover glass and another slide. Select a clean slide and put a drop of euparol. Now carfully place the coverglass on euparol. The mounting medium spreads to the edges of the coverglass. Extreme care should be taken to see that the side of the coverglass having material must face the slide and come into contact with euparol. Now take the original slide with the material, place a drop of euparol and gently slide a fresh clean cover glass. With this procedure, a single temporary slide can be made into two permanent slides. Dry the slides in a hot air oven till the cover glass does not move on the surface of the slide. Label them properly. Permanent slides are now ready for microscopic observation.

**Micrometry (Microscopic Measurements)**

This is a method employed to make measurements of microscopic objects like bacteria fungi, yeast cells, fibres, pollen grains, chromosomes etc. Two objects are necessary for microscopic measurements – an ocular micrometer and a stage micrometer. Ocular micrometer is a circular glass disc which has fine graduations engraved on one surface. The fine graduations can be seen only with the help of a microscope. The graduations engraved on the ocular micrometer are arbitrary though they are uniformly placed. Hence they have to be calibrated with the stage micrometer. A stage micrometer is essentially a glass slide with graduations. But these are definitely placed. Each line is 0.01 µm (10 micrometer) apart.

What is Calibration? An ocular micrometer by itself cannot be used for measurement because the lines are arbilary. Whereas in a stage micrometer the lines are 10 micrometers apart. In which case why use an ocular micrometer. One can use a stage micrometer placed on the stage for microscopic measurements. But if the stage of the microscope already has a stage micrometer it will not be possible to keep any other microscopic object (on a slide) for the purpose of observation. Hence the lines of ocular must be made to coincide with the stage micrometer and the distance between the lines of ocular known in terms of micrometers (µ) This is known as calibration.

After calibration there is no need to use the stage micrometer (provided one is using the same microscope and same magnification lenses).

How Calibration is done? The ocular micrometer disc is to be placed between the lenses of the eye piece. In order to do this remove the eye piece tube from the microscope, carefully unscrew the top lens, place the micrometer disc on a metal diaphragm below and screw the lens back. Replace the eyepiece tube on the microscope. In some microscopes the micrometer disc can be simply inserted at the bottom of the eye piece.

To calibrate the ocular micrometer for a given objective (45, 10, 60, or 100 x) it is necessary to super impose the scales of ocular and stage micrometers. After placing the ocular micrometer disc in the eye piece, place a stage micrometer on the stage and focus by using the coarse and fine adjustment knobs. Both the scales will be in sharp focus. Now carefully adjust the slide micrometer in such a way that the lines of both coincide as shown in the figure. Here seven ocular divisions match up with one stage micrometer division of 0.01 µm (10mm). In other words the value of 7 ocular divisions in 0.07/7 = 0.00143 mm. Since there are 1000 mm in 1 millimeter, each division in the ocular is 1.43 µm apart. After this value is known, the stage micrometer may be replaced with a slide of material to be measured. Calibration has to be done for each objective separately because the magnifications are different. Calibrate the ocular micrometer with the stage micrometer with 10 x 45 x 60 x and 100 x (oil immersion). A few examples of calibrating calculations are given below.

**Calculations :**

(a) 4 divisions of stage micrometer line up with 20 divisions of the ocular micrometer. What is the distance between each ocular division (OD) OD = 4/20 x 0.01 = 0.002 or 2 µm.

(b) 50 divisions of ocular are equal to 75 divisions of stage micrometer. What is the distance between each OD. OD = 75/50 x 0.01 = 0.014 or 15 µm.

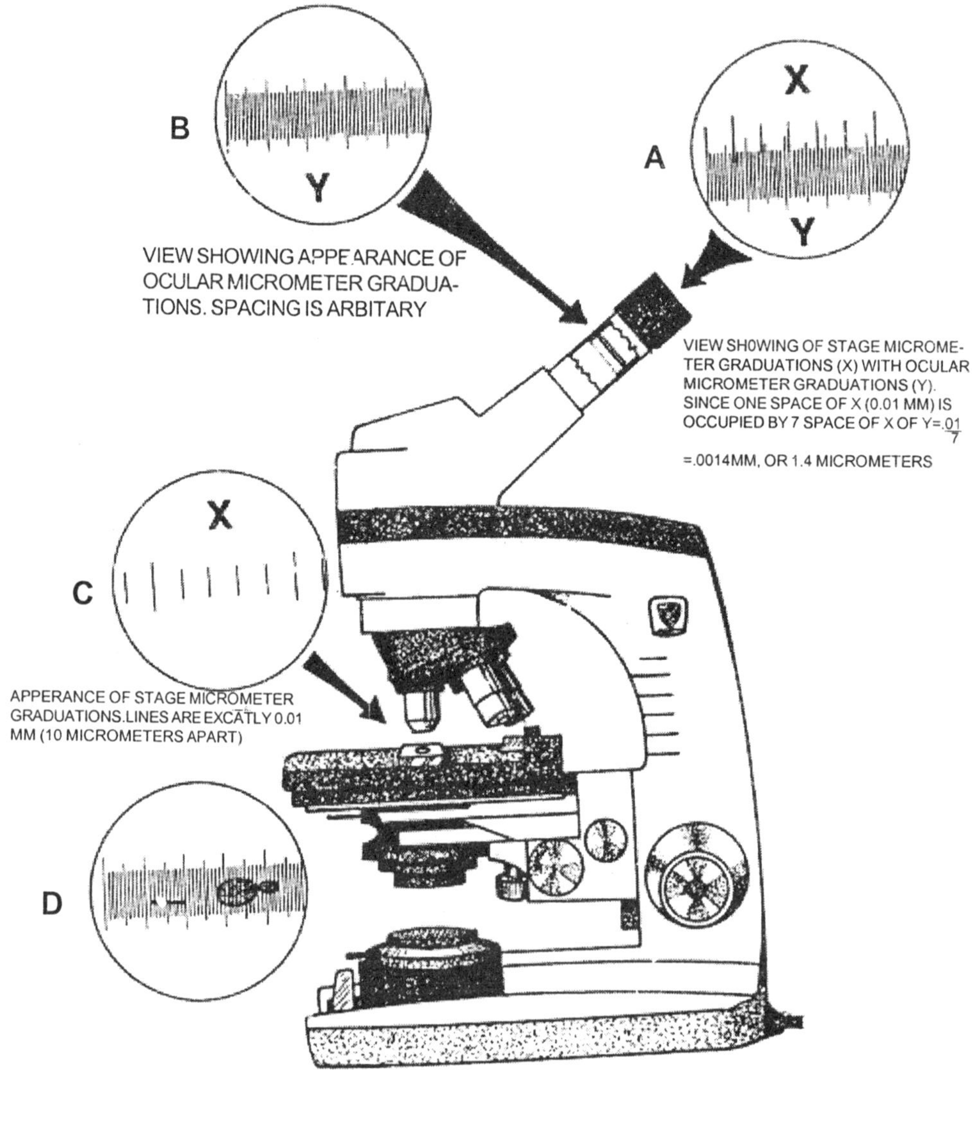


**Fig 15.1 Calibrations of ocular micrometer**

# MEIOSIS

Meiosis or reduction division occurs in higher plants and animals at the time of gametogenesis. Meiosis consists of the following stages. Meiosis I (First meiotic division) and meiosis II (Second meiotic division). Between the two is interkinesis.

**Meiosis I has the following stages**

**Interphase**

**Prophase I**

(a) Proleptolene (Proleptonema)
(b) Leptotene (Leptonenna)
(c) Zygotene (Zygonema)
(d) Pachytene (Pachynema)
(e) Diplotene (Diplonema)
(f) Diakinesis

2. Prometaphase
3. Metaphase I
4. Anaphase I
5. Telophase I.

**Meiosis II has the following stages.**

1. Prosphase II
2. Metaphase II
3. Anaphase II
4. Telophase II

The characteristic cytological features for the identification of these stages are given below.

**Interphase**

1. Nucleus appears to be resting, but it is highly active as it prepares the nucleus for division.
2. Synthesis of RNA and DNA will be maximum
3. Nuclear membrane intact, nucleolus can be seen clearly.

## PROPHASE

**Proleptotene**

1. Nucleus enlarges in size
2. Chromosomes are thin and hazy (except – for sex chromosomes)
3. Chromosomes appear double stranded.

**Leptotene**

1. Chromosomes are distinct, long and less coiled.
2. Chromosomes appear as threads with regularly placed beads or granules called chromomeres
3. The homologous chromosomes are associated at their tips where they are attached to the nuclear membrane
4. Leptotene chromosomes may be uniformly distributed or localized into boquets.
5. In some cases chromosomes may form tangled mass known as synzetic knot.
6. The centriole duplicates and each daughter centriole migrates to the opposite poles of the cell (in animal cells)

**Zygotene**

1. Chromosomes are shorter and thicker
2. Lengthwise pairing of the homologus chromosomes begins as they are attracted towards each other
3. Of the two homologues one is paternal and the other maternal. These are called bivalents.
4. Each homologus chromosome has two threads i.e. chromatids. In each bivalent there will be four chromatids.

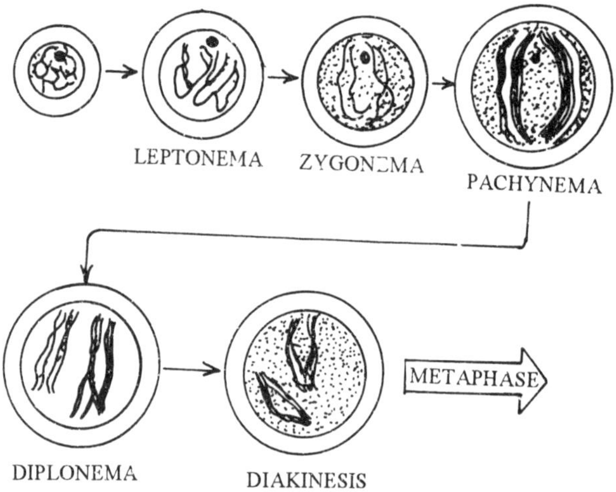


**Fig 15.2 Stages in Prophase I of meiosis**

**Pachytene**

1. Chromosomes of a bivalent are associated in tetrad (four chromatid) condition. Each chromatod pair is united by a centromere.
2. Due to longitudinal contraction pachytene chromosomes appear as thick threads.
3. The two bivalents twist around each other as relational coils (the coil of one twining with the other in opposite direction.
4. The chromatids also twist (coiling) resulting in tension.
5. Crossing over or chiasma takes place at this stage between non sister chromatids (opposing chromatids)
6. Two chromatids will be always out of perview of cross over.
7. Crossing over results in breakage and reunion of segments of non sister chromatids. After a break the same segments do not reunite. The chromatid which breaks at one point will unite with the segment of opposing nonsister chromated.
8. Crossing over may occur at 1, 2, or 3 points depending on the length of the bivalents.
9. Nucleolous remains unroganised. It may increase in size. It is found to be associated with certain chromosomes. These are called nucleolar organizing chromosomes.

(Note : Only details which can be seen under a light microscope are given. For example synaptonemal complex is organised at this stage. But this can be seen only under an electron microscope).

### Diplotene

1. Homologous chromosomes start separating (repulsion), but are held together at one or more points called Chiasmata
2. Chiasmata are clearer in diplotene than in pachytene because of increased condensation of chromosomes.
3. Chiasma may be formed at the ends (terminal chiasma) or in the middle (interstitial chiasma) of the chromosomes.
4. Terminalization or uncoupling of chromosomes begins due to repulsion between homologous chromosomes.

### Diakinesis

1. This is the final stage of prophase I.
2. Bivalent chromosomes are very thick and lightly contracted.
3. Bivalents are distributed more towards the periphery of the nucleus.
4. Terminal chiasma may be retained maintaining contact between homologous chromosomes.
5. Nucleolous detaches from the chromosome and finally disappears.

### Prometaphase I

1. Nuclear membrane disappears.
2. Centrioles are joined by spindles which develop from tubules (in animal cells)
3. Chromosomes reach their maximum contraction

### Metaphase I

1. Spindle becomes clearly visible and becomes more pronounced
2. Bivalent chromosomes arrange themselves on the equatorial plate.
3. Centromeres of the bivalents lie towards the poles while the arms lie towards the equator
4. Centromere is attached to the spindle fibres. These are called chromosomal fibres.

### Anaphase I

1. Each homologous chromosome of each bivalent moves towards the opposite pole.
2. Homologues do not separate in all instances immediately, while shorter chromosomes can be seen separated, longer chromosomes will still be in contact. (they take a longer time to separate).
3. Chromosomes assume 'V' shape as they are pulled towards the opposite poles.
4. The centromere does not divide.

### Telophase I

1. Chromosomes reach respective poles, spindles persist for some time
2. Chromosomes undergo despiralization, becoming indistinct
3. Nucleolous reappears, so also the nuclear membrane. In some cases nucleolous may not reappear at this stage.
4. Cell wall is formed in plants in the form of a plate. In animals cells cell membrane is formed by furrowing.
5. In some of the plant cells cytokinesis will not take place at this stage.

### Interkinesis

1. The daughter nuclei or daughter cells undergo a resting period
2. It is of a very short duration

### Meiosis II

Second meiotic division is essentially a mitotic division in which both the chromosomes and cell divide.

### Prophase II

1. Centrioles divide into two and two pairs of centrioles are formed. They migrate to the opposite poles (in animal cells)

2. The microtubules arrange themselves at right angles to the previous plane.
3. The arms of the chromosome remain separate and there is no relational coiling.
4. Nucleolous as well as nuclear envelop disappear with the formation of achromatic figure.

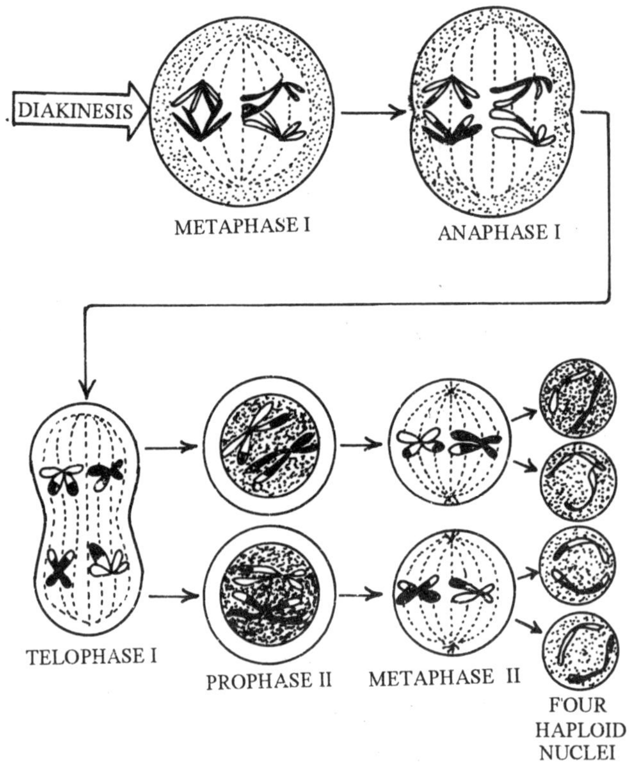


**Fig 15.3 Stages in meiosis (after prophase)**

**Metaphase II**

1. It is of very short duration
2. Chromosomes with their chromatids arrange themselves on the equatorial plane.
3. The centromeres lie on the equator, while the arms are directed towards the poles.

4. The centromere divides, thus separating the two chromatids which eventually become separate chromosomes
5. Centromere is attached to the chromosomal spindle fibres in the achromatic figure.

**Anaphase II**

1. The daughter chromosomes start their migration towards the opposite poles.
2. Chromosomes are not very thick and short like in analphas I, but are like the anaphase daughter chromosome of mitosis.

**Telophase II**

1. The chromosomes having reached the opposite poles uncoil and become less distinct.
2. Nuclear membrane and nucleolous reappear resulting in the information of a daughter nucleus.
3. The nucleus gets into the interphase state.
4. The nuclei are haploid.

**Cytokinesis**

1. Wall formation begins either as a furrow or as a plate and two daughter cells are formed.
2. At the end of meiosis four daughter cells are formed.
3. In some cells cell wall would not have been formed after meiosis I as a result four nuclei are formed. Cell walls are now laid down resulting in the formation of four cells.

**MITOSIS**

Mitosis occurs in the somatic cells of multicellular organisms and it is the chief means of multiplication of cells contributing to the increase in size oft the individual.

For the convenience of study, mitosis may be divided into two stages – karyokinesis and cytokinesis. Karyokinesis is again divided into the following phases – a) Interphase, b) prophase c) metaphase d) anaphase and e) telophase. The following are the characteristics of these stages.

**Interphase**

1. No morphological changes are visible in the nucleus.
2. It was originally thought to be a resting phase between two nuclear divisions.
3. Recent studies have indicated that interphase is the most active as thesynthesis of RNA and DNA takes place in order to prepare the cell for division
4. The interphase nucleus is large and conspicuous, so also the nucleolus. (Details of G, S and $S_2$ phase are not given as they are theoretical observations)

**Prophase**

1. Cells become spheroidal with increase in its viscosity and refractivity (in animal cells)
2. Chromonema starts condensing and ultimately towards the end of prophase thick threads called chromosomes are formed. Chromosomes contact up to $1/25^{th}$ of their original length. Specific number of chromosomes are formed.
3. Each chromosome longitudinally splits into two identical chromatids connected by a centromere
4. The two chromatids are coiled round each other and cannot be easily separated
5. Nuclear matrix begins to accumulate on the chromatids and the chromosomes assume a hazy outline.
6. Centrioles which have undergone duplication during interphase and situated above the nucleus start moving in opposite directions (in animal cells)
7. Astral rays and spindle fibres (amphiaster) appear. But amphiaster is not visible at this stage as it does not take stain (in animal cells)
8. Nucleolous decreases in size and finally disappears.
9. Towards the end of prophase, nuclear membrane disappears.

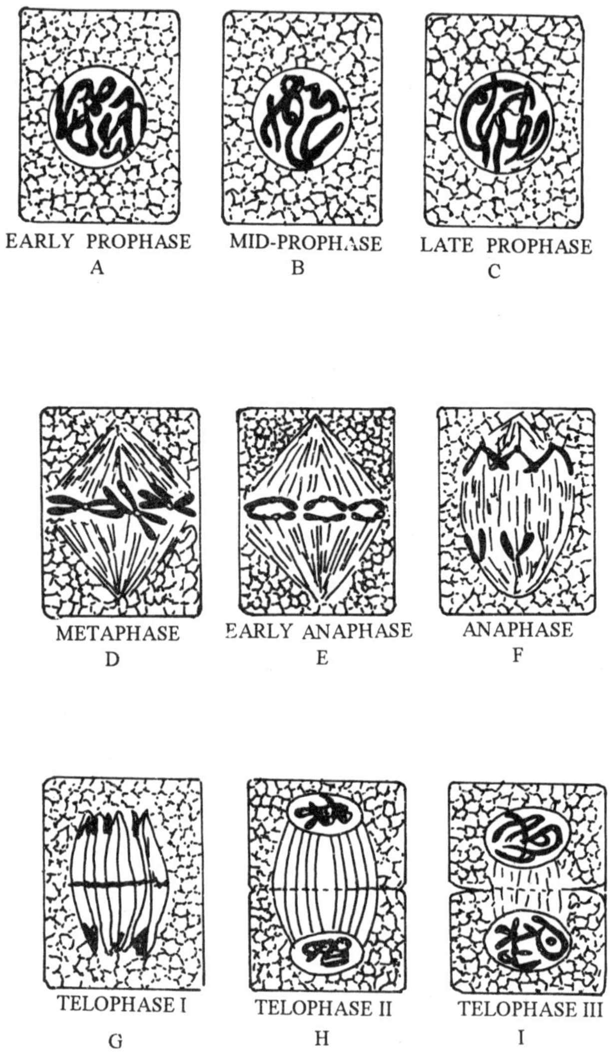


**Fig 15.4 Stages of mitosis in a plant cell**

**Prometaphase**

1. Nuclear envelop totally gets disentegrated

**Metaphase**

1. Duration is short
2. In early metaphase highly condensed and shortened chromosomes arrange themselves radially on the equatorial plane. It is called the metaphase plate or the equatorial plate
3. Chromosomes lie at the equator with the centrioles pointing towards the poles attached to the spindle fibres (in animal cells)
4. In plant cells chromosomes are irregularly scattered on the equatorial plate. Smaller chromosomes lie at

the centre, larger chromosomes towards the peripherey.

5. Chromatids become separated from each other
6. Centromere divides so that the chromatids get separated each having its own centromere.

**Anaphase**

1. After the division of the centromere, the original number of chromosomes gets doubled
2. Exactly half of them will begin to migrate to one pole and the other half to the other pole.
3. Spindle fibres are distinct at this stage. They are of three kinds (a) Chromosomal spindle fibres attached to the chromosomes, (b) Spindle fibres attached to the asters known as continuous spindle fibres and (c) spindle fibres between departing chromosomes called interzonal spindle fibres (in animal cells)
4. In plants all the spindle fibres are said to be anastral as they are formed without centrioles or asters.
5. During the polar migration chromosomes appear as rod shaped 'L' shaped or V shaped 'L' shaped chromosomes are called heterobrachial (unequal arms) while V shaped chromosomes are called homobrachial (in equal arms).
6. The some instances some chromosomes do not move and remain at the equatorial plate only. They are called laggards and are without centromere.
7. Usually sex chromosomes migrate slowly (generally found in animal cells)
8. At the end of anaphase clusters of chromosomes appear at each pole of the cell.

**Telophase**

1. Chromosomes begin to uncoil into their strands and become distinct
2. Nucleolus begins to appear as the nucleonemal strands become thicker
3. Spindles disappear
4. Asters lose their fibres and are reconverted to centrioles (in animal cells)
5. Nuclear membrane reappears forming a regular nucleus.

**Cytokinesis**

1. In plants cell plate is formed between two groups of chromosomes
2. Cell plate is formed at the centre and growing on either side joins the cell wall.
3. In animal cells cytokinesis is by furrowing.